l AL MARKETING

THE RIGHT MARKETING AT THE RIGHT TIME

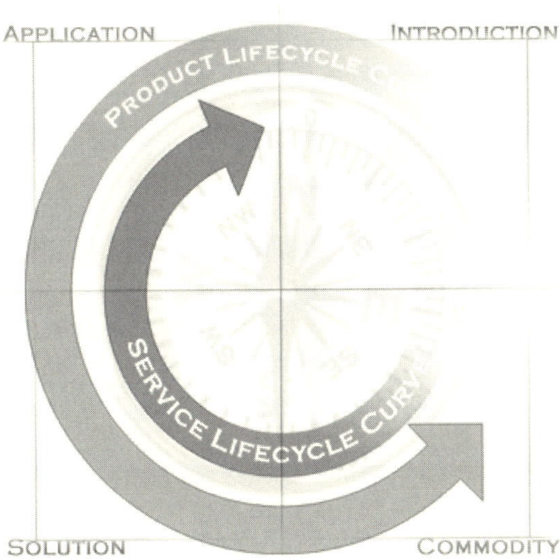

Featuring the Market Maturity Matrix™

Patrick Perugini

with Jeff Wiss

For more information or to purchase additional copies of this book for educational, business, or sales promotional use, please write to Richards-Stallman Publishing Company, 295 N. Cherrywood Drive, Lafayette, Colorado, 80026, or visit www.richards-stallman.com. To contact the author by email, write to patrick@maccelgroup.com.

Cover design by JB Leach, kubicos@home.com

ISBN 0-9715897-0-4

To my wife Debby
whose love and laughter
are an eternal source of
joy and encouragement.

Contents

Acknowledgments

Completing this book has been the highlight of my last 14 years in technology marketing. During this time, I have had the pleasure of working with many great individuals in some dynamic companies. Although the list of these individuals is too long to include here, I want to recognize them for the mark they left on my career. This book stands as a testament to their intellect, patience, and guidance.

As a struggling young marketer in the late 1980s, I made my share of mistakes - as everyone did - trying to apply "old school" or consumer marketing strategies and tactics to technology products and services. Not surprisingly, results were unpredictable and difficult to repeat. Worse yet, with each successive company I worked with the situation changed and the techniques I had always counted on no longer applied.

What I realized I needed was a methodology - not just marketing theory or guerrilla tactics - but a reliable approach to technology marketing that made results more predictable and credible. So this became my quest.

Although I have been practicing the fundamental concepts of this book for years, it wasn't until I co-founded Maccel Group in mid-2000 that I realized I wasn't alone in my quest. Every marketer I met shared the same frustration with currently available approaches to technology marketing. Every executive I met felt frustrated that no two marketers approached marketing strategy or tactical planning the same way.

In listening to these companies and individuals, I realized the importance of writing this book. It is the culmination of over a year of research, writing, and editing, and it would not have been possible to complete without the generous assistance and involvement of many people.

First, I would like to thank Jeff Wiss for his unique insight and expertise that helped refine the Market Maturity Matrix™. Our success in applying the principles of this book on over 100 technology companies has served as both the inspiration and validation of this book.

Next, I would like to thank John Metzger, CEO of Metzger Associates for contributing his time, energy, and passion to this book. John was one of the first people to recognize the potential for the Market Maturity Matrix, and provided a great source of encouragement to write this book. Because of this, he was a natural choice to write the foreword.

I would also like to thank the technology companies that have applied and adopted the Market Maturity Matrix to their business. Our experience with these outstanding companies has helped to refine the model, and aided in the development of the best practices discussed throughout this book.

In addition, I would like to thank those technology service providers including advertising agencies, public relations firms, web and media design professionals, sales consultants, venture capital firms, and technology incubators that have made the Market Maturity Matrix an integral part of their business. In particular, I want to recognize Metzger Associates Public Relations of Boulder, Colorado, and Cactus Marketing Communications of Denver, Colorado. These two companies, along with their exceptional employees and clients were the earliest adopters of the Market Maturity Matrix and the *Perpetual Marketing* approach. Through their pioneering efforts, they proved that the Market Maturity Matrix is the fastest way to turn strategic decisions into predictable results.

I would also like to thank Sharon Brant, James Clark, Mitchell Derman, Steve Dewitt, and Karen Scofield for their contributions to this book.

Finally, I would like to thank our wives, Debby Perugini and Diane Wiss, who continue to provide Jeff and I with more support, love, and understanding than we deserve.

-- *Patrick Perugini*

Foreword

By John Metzger

During the mid-20th Century, forward-thinking marketers developed bold strategies to sell innovative, breakthrough consumer products like laundry detergent, paper towels, and potato chips. They understood their customers, and knew exactly how to reach them with the right messages, in the right places, at the right times. Once set in place, the communications strategies for these brands and products remained relevant for decades.

Nearly half a century later, marketers continue to apply the same techniques to sell microchips, software, cell phones, and other technology products and services. Sure, it still can work. But only sometimes - and only if you're lucky. Market adoption rates, economic conditions, and technology itself evolve much faster and more dramatically than they do for packaged goods. Yet many cutting-edge companies keep developing strategies that take far too long to conceive, and become completely obsolete by the time the tactics to support them are implemented.

After founding one of the nation's first high-technology public relations firms in 1991, my clients and I were thwarted by "strategy" that failed to accommodate paradigm shifts. I was frustrated by CEOs whose unrealistic marketing goals set up their communications programs and companies for failure. I was baffled with delusional marketing that positioned products and services in the wrong way, at the wrong time to the wrong audience.

I wanted a methodology to serve as a "watch" and "compass" that would tell us where our clients were in terms of market timing, buyer behavior, and adoption rates. I wanted everyone - from our clients' executives to our youngest story pitcher - to understand where we were, where we were going, and how we were going to get there. I wanted a methodology that helped us monitor marketplace changes and watch for the slightest changes in the competitive environment, early adoption into new vertical markets, key mergers and acquisitions, and other factors so we could adjust our tactics on the fly. I wanted our clients and PR teams to sing off the same songsheet, with full consensus on what we were doing and why. I wanted our messages to resonate consistently across the company, from the receptionist to sales, and from engineering to marketing.

The theory brought forth in Geoffrey Moore's book, "Crossing the Chasm," started to "get it." Moore identified correctly the concept that all forms of innovative products and services require "early adopters" to be the first to catch on and take the lead – a process unique to technology markets. If the early adopters crossed the chasm, others would follow. But in reality, technology markets, products, and services cross many chasms over their lifecycle. And whenever you think you've crossed them all, you'd better search out another chasm to cross to maintain a leadership position.

We found our answer in the Market Maturity Matrix developed by Patrick Perugini. For the first time, marketers of technology products and services have a methodology that specifically works for them, and allows them to truly do "the right marketing at the right time." At Metzger Associates, we have adopted the Market Maturity Matrix as our strategic platform for all our clients. We've even taken it a step further and developed our own public relations "overlays" that outline public and investor relations best practices that are appropriate for each market phase. We also work with and recommend service partners who have adopted the Market Maturity Matrix as we have. This saves our clients a lot of time and trouble by having their PR, IR, ad agencies, web designers, sales teams, and others using the same platform, rather than each one advocating their own.

The Market Maturity Matrix is a methodology that's understood not just by our highest level strategists, but by all our clients, at all levels, and by everyone on the PR teams that serve every client. It gives us a language to speak. A process to understand. Each individual on each team can give solid advice, with the utmost conviction, on how our clients communicate to multiple audiences during various stages of market upheaval and change.

I have waited my entire career to find this solution. Metzger Associates is an "early adopter," and we apply it to our specialty of public relations and investor relations. But I'm confident that *Perpetual Marketing* will spread throughout the entire technology industry. It can help lead us into the brave new world of technology product and service marketing and the tremendous discoveries, productivity gains, and market expansion that they offer. It is my distinct honor to kick off the revolution that Patrick and Jeff have started. Use this book at first to understand how the Market Maturity Matrix can make your company more efficient and successful. Once you "get it," we can all work together to change the world.

Introduction

It's often said that the only constant in business is change. Nowhere is this truer than in technology markets. This is because technology markets are in a constant state of motion and transition - a perpetual motion - driven by a wide variety of technological, competitive, economic, and regulatory forces. From innovation to commodity, technology markets are influenced by seemingly unpredictable changes in buyer behavior, unexpected leaps in technology, unforeseen competitors, and a never-ending stream of mergers, acquisitions, public offerings, and failures. It's no surprise that companies considered industry leaders one day are ancient history the next.

Remember Lotus 1-2-3? How about Harvard Graphics? Remember Ashton-Tate and dBase? Novell NetWare? How about the Hayes Modem? What about dongles? Or dot-matrix printers? All these were once considered the industry standards by which other products and technologies were compared. Many were considered unstoppable. But as the markets for these products and technologies changed, these companies did not change with them, and found themselves unable to compete under the new market rules.

This phenomenon will forever be a part of the technology marketplace. Someday, we'll likely look back and reflect nostalgically about CRT monitors, mice, floppy disk drives, keyboards, cellular phones, and probably many of the technology companies we once worked for.

The lesson here is that change is inevitable, but it isn't the enemy. Change offers us a choice. It can either be a source of eternal frustration or perpetual inspiration. The imperative and point of this book is to capitalize on market transitions and maximize the opportunities at every market stage.

Why You Need This Book

Entering and competing in technology markets is becoming increasingly expensive. For this reason, technology marketers have a responsibility to be more disciplined in their approach to budgeting, planning and execution. They can't afford to waste time, money, and resources on efforts that don't yield the greatest return on investment. This was a lesson learned hard and fast by countless dot-coms as they raced to market using marketing for brand building and hypermarket growth rather than to drive business strategy, focus, and prioritization.

To succeed in today's markets, technology marketers need a proven, predictable, and credible approach to market strategy and tactical planning. This not only minimizes the risk of failure, but it enables marketing to play a more central role in this new age of strategic corporate planning.

Unfortunately, most current marketing theories and approaches don't help. They aren't reliable for understanding and predicting the technology markets across every market stage. They are often based on traditional markets where market dynamics changed over decades, not days. Or they are based on consumer adoption habits, which aren't easily applied to business-to-business markets. In technology markets, where market transitions, economic cycles, innovations, and buyer behaviors can change overnight, technology marketers need a methodology unique to technology markets. More importantly, they need one that focuses and prioritizes resources and efforts to do "the right marketing at the right time."

This book will introduce you to an innovative approach to technology marketing using a methodology called the Market Maturity Matrix™. This methodology offers technology businesses a more disciplined approach to marketing technology products and services that yield better results, faster. The Market Maturity Matrix does this by using a graphical model to illustrate how product and service markets mature and evolve, and how marketing strategy and tactics must adapt to changing market conditions. It also uses buyer behavior, competitive dynamics, and key market indicators to divide market lifecycles into distinct and predictable phases, each with its own unique characteristics.

However, the Market Maturity Matrix goes beyond marketing theory. It uses these market phases to identify marketing tactics proven to yield the greatest return at each stage of the market.

How to Use This Book

This book is divided into three sections:

- Section I includes an in-depth discussion of *Perpetual Marketing* and the Market Maturity Matrix. This includes descriptions of technology product and service market lifecycles, market dynamics for each market phase, and case studies.
- Section II includes chapters on how to interpret and apply the Market Maturity Matrix, including market transitions, market entry strategies, buyer behaviors, the "Laws of Perpetual Marketing," and key marketing tactics at each market stage.
- Section III includes a self-assessment questionnaire to help you quickly assess your market's maturity on the Market Maturity Matrix model.

Become a Perpetual Marketer

With this book, we will bring renewed confidence and credibility to your marketing strategy, tactics, and budgets. We will also make you more persuasive, insightful, and impactful in your discussions with other executives by introducing a methodology that can become the cornerstone of your business planning process. By the time you finish reading this book, you will be able to answer these important questions:

- How can I better prepare for and respond to market changes and transitions?
- How can I be sure I am doing the right marketing for my business right now?
- What is the best strategy for entering the market right now?
- How can I make my marketing plans more predictable, credible, and fundable for executive management?
- How can I help executive management better understand and appreciate marketing-led business strategy?
- How can I better focus and prioritize marketing resources in a seemingly unpredictable industry?
- How can marketing play a more influential role in business decision making?

Section I:
Introducing Perpetual Marketing

1

Markets Change, So Why Don't We?

Technology companies often ask us, "Why isn't our marketing working anymore? It's always worked before."

Our response is, "What's changed?"

The answer is usually that the market has matured but the company has not. In technology markets, conditions change so fast that what worked before, may not work now. The right marketing today, may be very different tomorrow.

So, when you wonder why something doesn't work the way it used to, first ask yourself, "What's changed?"

Great Marketing, Now Change It

Albert Einstein once defined insanity as, "doing the same thing over and over again, and expecting different results." So if technology markets are in a constant state of motion and transition, it stands to reason that our approach to marketing must also transition to match the velocity of the market's change. However, when the velocity of change can vary significantly from market to market, it is difficult to estimate and predict the "clockspeed" of a market - the speed and direction in which a market matures. Despite this, technology companies make strategic decisions every day based on inconclusive data about their market's future. Where will the market be next month? Next year? In three years? Who will the competition be? What sales and delivery channels will be used? How will our buyers change?

These are critical questions when you consider that historically the maturity rate of technology markets has been increasing exponentially. For instance, it has been documented that it took 38 years for radio to reach 50 million listeners. It took television only 13 years to reach the same number of viewers. The Internet reached 50 million users in only 4 years. Napster went from zero to 50 million users in only six months!

Because technology markets can change so rapidly, marketers need to be near-sighted, far-sighted, and hind-sighted - all at the same time. They have to know how to prepare for, respond to, and leverage changing market opportunities with the right strategy and tactics at the right time.

To do this, technology marketers need to track and predict market changes throughout their market's evolution and understand market dynamics such as:

- Key indicators that predict changes in a market
- Changing buyer needs across market stages
- The impact of economic and regulatory forces
- The effect of competitive and complementary technologies
- Market clockspeed and direction
- Converging and diverging markets
- Market segmentation and expansion

Such understanding helps:

- Predict and maximize market transitions
- Respond to evolving buyer characteristics
- Adjust marketing strategy, budgets, and resources to accommodate changing market stages
- Stay one step ahead of the competition
- Capture and maintain market share
- Select and target markets
- Become the first mover in your market with or without technology leadership

Introducing Perpetual Marketing

Technology markets are like objects in motion. When forces influence a moving object, they impact the object's speed, direction, and velocity in a predictable fashion. Similarly, as forces act upon technology markets - such as innovations, buyer behaviors, competition, and economic changes - they have predictable impacts on market direction and speed.

Because of these similarities, the study of technology markets is comparable to the study of motion - in particular the theory of Perpetual Motion. In fact, by applying the principles of Perpetual Motion to technology markets, businesses can better understand and predict market behavior, and the likely effect forces have on markets at different market phases. This is the essence of *Perpetual Marketing*. It allows companies to use a more scientific approach to strategic decision-making and tactical market planning.

The Right Marketing at The Right Time

In technology marketing, there is a right time and a wrong time for applying marketing resources and activities. The right marketing at the right time can speed adoption rates, shorten sales cycles, and target appropriate prospects, channels and partners. Conversely, the wrong marketing at the wrong time can be fatal for technology businesses.

***Law of Perpetual Marketing**: Perpetual Marketing is simply "the right marketing at the right time."*

With the steadily increasing cost to enter and compete in technology markets, companies can no longer afford to experiment with precious marketing time and resources. We have all heard the stories of the technology companies who advertised at the 1999 Super Bowl, only to be out of business the following year. These were indicative of costly strategic marketing mistakes. However, mistakes in marketing strategy and tactics are usually not this obvious - but with equally damaging results. With *Perpetual Marketing*, you will learn how, why and when to apply the right marketing to avoid common pitfalls, and maximize marketing resources. Perhaps more importantly, it guides companies on what *not* to do.

The bottom line is: *Perpetual Marketing* offers technology product and service companies a more disciplined approach to making strategic market planning decisions, while providing marketers a new degree of credibility in developing marketing plans and budgets.

The Market Maturity Matrix™

At the heart of *Perpetual Marketing* is the Market Maturity Matrix™, a unique methodology designed exclusively for technology marketing. The Market Maturity Matrix combines market strategy and tactics into a single model based on the lifecycle speed and direction of technology product and service markets.

Law of Perpetual Marketing: *The Market Maturity Matrix acts as both a compass and a timepiece for technology markets.*

Unlike traditional bell curves used to describe market adoption distribution, the Market Maturity Matrix uses buyer behavior, competitive dynamics, and key market indicators to chart market lifecycles, track market progress, predict market transitions, and distinguish market characteristics across evolutionary phases. This makes the Market Maturity Matrix the single most effective strategic planning model available to technology marketers.

Law of Perpetual Marketing*:* *Technology markets are distinct and predictable across market stages.*

However, the Market Maturity Matrix offers more than just strategy and theory. Perhaps more importantly, it uses the distinct characteristics of each market phase to identify unique market opportunities, and align best practice tactics with market strategy.

Using this proven approach, marketers and executives can navigate technology markets with greater confidence and credibility. It also allows businesses to predict and respond more effectively to market changes by continually adjusting and refining their marketing, sales, and operations to capitalize on changing market and buyer needs.

For marketers, the Market Maturity Matrix helps make critical decisions related to:

- Market selection and expansion
- Partner and channel planning
- Media and press relations
- Direct and indirect market planning
- Market entry
- Customer retention

- Positioning and messaging
- Pricing and promotions
- Selling systems

For business planners and executives, this insight extends into:

- Sales force planning and forecasting
- Product development
- Market sizing
- Financial planning
- Raising capital
- Recruiting
- Merger and acquisition planning
- Infrastructure planning

Bringing Methodology to the Madness

Many executive teams have traditionally considered marketing a "soft" discipline. They have historically found marketing plans to be inherently suspect because they are based more on instinct and personal experience than on "science."

In contrast, sales, development, support, and accounting organizations all use standardized processes, tools, and techniques for planning and measurement based on industry best practices. For instance, the sales department may follow a sales process or methodology such as Solution Selling, Strategic Selling, or Spin Selling. Engineering departments typically follow a structured and repeatable process for product development and delivery. Accounting follows GAAP, the Generally Accepted Accounting Principles. As a result, these processes not only accelerate their decision-making, but also improve the predictability and consistency of their results.

Unfortunately, even when companies recognize the need for a more scientific or disciplined approach to marketing, it often has difficulty supporting this with a proven and repeatable methodology. Now, with the Market Maturity Matrix, such a methodology is finally available.

When first introduced to executives, the Market Maturity Matrix provides an "ah-ha" moment. This is because for most executives, it is the first time many of them have been exposed to a marketing tool that truly reflects their business and market and offers practical, credible insight for strategic planning. Using the Market Maturity Matrix, executives now know why their markets do what they do and how they can leverage this to create and sustain competitive advantage.

2

The Significance Of Market Maturity

To determine the market maturity for a client's products and services, we start by asking a series of poignant questions about the behavior of their buyers, their competition, and the applications for their products and services.

Too often the answers are, "We don't know."

Many companies do not understand that technology markets are always in motion. They fail to recognize the importance of always knowing their market's maturity in terms of buyer behavior, key market indicators, channel maturity, competitors, applications, and business drivers. Market maturity establishes the starting point to help determine what the right marketing is at the right time.

The reason many companies today don't understand their market's maturity is because they don't think in terms of markets – they think in terms of their business. In turn, most companies approach market strategy based on the maturity of their company, product or service; not the maturity of the market in which they compete.

For instance, if a start-up company is preparing to launch its first product, how do they determine what marketing activities will yield the best results? How would they segment and target the market? How do they prioritize channels and partners? How do they align their positioning and messaging to the behavior of buyers at this market stage? How do they disrupt the competition? How do they prepare for inevitable market transitions?

These are difficult questions to answer if company strategy is based on the company's maturity versus the market's maturity.

The better approach is to first understand the market's maturity. This tells you much more about what buyer characteristics are like, what your competition is likely to do next, and what market segments and levels of domain or industry expertise are necessary in your sales and marketing efforts. Approached this way, it is much easier to determine the business and marketing practices that will yield the best results.

Creating the Market Maturity Matrix Model

Technology markets are constantly being pushed, pulled, and dragged by a never-ending supply of technological, economic, environmental, and regulatory forces. However, of all the forces that impact technology markets, the two most influential are Technology Evolution and Market Adoption. For this reason, these two forces are used as the two axes of the Market Maturity Matrix model, as illustrated in Diagram A.

***Law of Perpetual Marketing**: The two most influential forces driving technology markets are technology evolution and market adoption.*

Diagram A. Market Maturity Matrix Axes and Quadrants

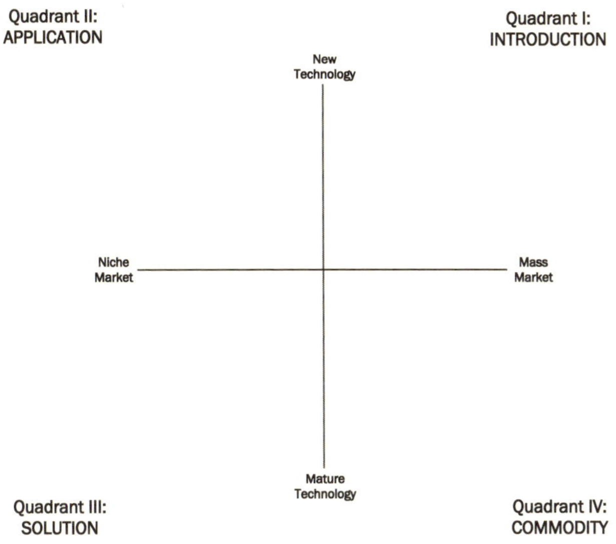

Technology Evolution – from new to mature technology – is the vertical axis. Market Adoption – from niche to mass markets – is the horizontal axis. When aligned against one another, the intersection of these two forces creates a model with four distinct quadrants, labeled counterclockwise in Diagram A starting in the upper right quadrant as Introduction, Application, Solution, and Commodity.

Technology Evolution

Technology evolution follows the maturity of technology from innovation to maturity. Starting with discontinuous innovations, technology evolution moves down along the vertical axis of the model. Along the way it is aided by a number of continuous innovations until it reaches maturity.

For the purposes of interpreting the Market Maturity Matrix, it is important to understand the distinction between discontinuous innovations and continuous innovations.

***Law of Perpetual Marketing**: Technology innovations can be continuous or discontinuous. Know the difference.*

Discontinuous Innovations

Herman Melville once said, "It is better to fail in originality than succeed in imitation." Such is the inspiration for discontinuous innovations - the market disrupters that create "paradigm shifts" in technology or introduce completely new ways of solving problems. Discontinuous innovations are the start of technology product lifecycles. They promise vast improvements in speed, quality, or cost savings. At the same time, discontinuous innovations often come with significant adoption or replacement costs.

Video Cassette Recorders (VCRs) are a great example of a discontinuous innovation that created fundamental changes to television viewing habits. Prior to the development of the VCR, there was no way to tape television programs at home. So when the earliest manufacturers introduced products to market, they were able to charge early adopters a premium price for this new technology. Early VCRs ranged between $500 and $1,000. Competing formats including Beta (primarily backed by SONY) and VHS (backed by nearly everyone else) targeted different types of customers, and helped to maintain elevated prices by preventing the market from reaching standardization. It wasn't until the market ultimately adopted VHS as the standard that prices came down, making VCRs affordable to mass markets.

Continuous Innovations

Continuous innovations are the incremental enhancements that build on an existing technology platform with features, functionality, ease of use, performance, and other enhancements. These are typically intended to extend the use, longevity, application, or compatibility of these existing technologies.

Microsoft Windows is a classic example of continuous innovation. Version 1 of Windows was primarily designed as a graphical user interface for the underlying operating system, DOS. Over the years, subsequent versions of Windows added significant capabilities that allowed it to continually extend its control of the desktop, user interface, hardware, and peripherals, and ultimately eliminate the underlying DOS components.

Market Adoption

Market adoption distinguishes the way in which markets embrace products and services. Adoption rates are impacted not only by buyer characteristics, but also by industry trends, key market influencers, and economic and regulatory conditions. On the horizontal axis of the model, the two extremes for market adoption are niche markets and mass markets.

As you will see, niche and mass markets have different expectations and acceptance levels to new technologies. As a result, understanding the market adoption of your product or service will have dramatic implications on marketing decisions including how products and services are packaged, priced, positioned, marketed, and delivered to customers at each stage of a market's maturity.

For instance, the Market Maturity Matrix teaches us that mass markets buy more tactically. Niche markets buy more strategically. So, the broader the market or more general the buyer's need, the more "packaged" or standardized a product tends to be. The narrower the market or more unique the purchaser's needs, the more "solution-oriented" or customized a product tends to be.

***Law of Perpetual Marketing**: Mass markets buy tactically. Niche markets buy strategically.*

Examples of tactical purchases include disk drives, laser printers, and tax preparation software. These commodity market purchases are used to address very specific needs, and can be bought through mass merchants. Strategic purchases include Knowledge Management Software, 3D graphic animation software for the web, or network appliances. These solutions are being used as change agents or to create competitive differentiation.

Market Lifecycles

History and experience have taught us that product and service lifecycles are inherently different. In turn, they follow different evolutionary paths. History has also taught us that product and service lifecycles are predictable in how they respond to Technology Evolution and Market Adoption.

Diagram B illustrates how product and service market lifecycles apply to the Market Maturity Matrix model.

Diagram B: Market Lifecycles

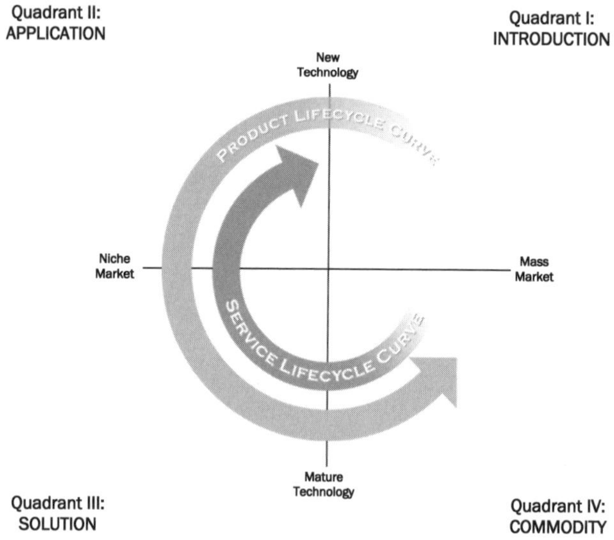

The outer curve represents product lifecycles, moving counterclockwise from the upper right quadrant (Introduction) to the lower right quadrant (Commodity). The inner curve represents service lifecycles, moving clockwise from the lower right quadrant (Commodity) to the upper left quadrant (Application). Each of these quadrants represents a distinct phase in market lifecycles.

Key Market Indicators

One of the ways we track product and service lifecycles on the Market Maturity Matrix is by using Key Market Indicators. Each key market indicator represents an important evolutionary milestone that companies can use to identify market maturity, measure market progress, and predict market transitions. Table A identifies some of these key market indicators.

Table A: Key Market Indicators

Product Lifecycles	Service Lifecycles
Innovation	Service Business Model
Competition	Competition
Market Segmentation	Service Methodology
Market Consolidation	Service Solution
Product Solution	Service Bundling & Packaging
Channel Maturity	Sell Product Innovation
Technology Standardization	
Embedded Technology	

On the Market Maturity Matrix, these key market indicators are represented as points along the market lifecycle curves, as seen in Diagram C.

Diagram C: Market Lifecycles with Key Market Indicators

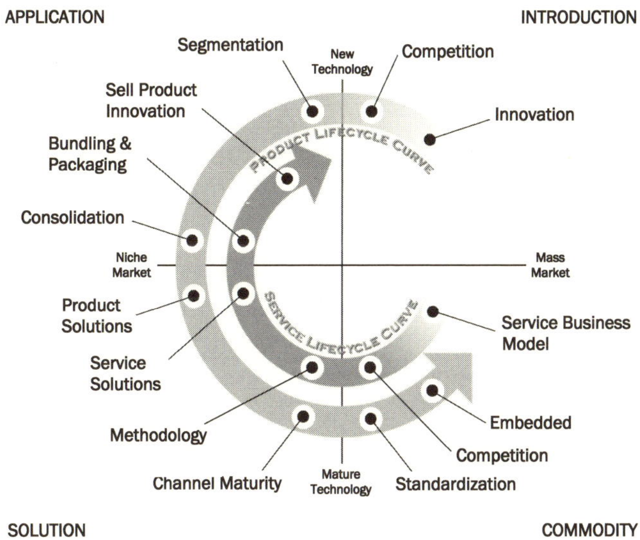

Using the Market Maturity Matrix, theses key market indicators represent predictable milestones for market evolution. They help to track market direction and speed, and identify market phase transitions.

3

Market Lifecycles At-A-Glance

Clients often ask us, "Why do product markets move in one direction on the model while service markets move in the other?"

Without oversimplifying it, the short answer is, "Product and service lifecycles differ primarily due to the effect technology innovation has on each market. Product lifecycles are the source of technology innovation, taking it from introduction to commodity. Service lifecycles start with commodity technology and introduce it to entirely new markets by focusing not on technology, but on delivery and management."

This chapter offers brief introductions to technology market lifecycles in advance of the more detailed chapters to come. The purpose of this chapter is to give you an overview of product and service market lifecycles, present case studies for each, and establish the interrelationship between these two markets.

PRODUCT LIFECYCLE CURVE

Product Lifecycles evolve counterclockwise on the Market Maturity Matrix from the upper right quadrant – the Introduction quadrant – to the lower right quadrant – the Commodity quadrant – as illustrated in Diagram D.

Diagram D: Product Lifecycles

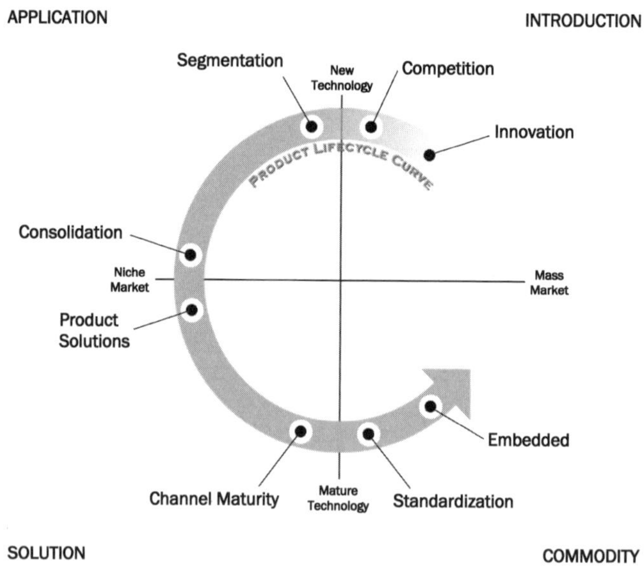

Phase I Product lifecycles start as discontinuous *Innovations*. However, since these Innovations frequently originate with more emphasis on technological potential than on market potential, discontinuous innovations often overestimate market opportunity and are initially introduced to large, unqualified audiences. Hence, these innovations are introduced to mass markets like shouts from a rooftop, hoping to reach anybody eager to be the first to try out new technology.

Once these innovations are introduced into the market, *Competition* – both real and perceived – emerges quickly for fear that they will miss out on "the next big thing." However, since mass markets are generally too expensive to launch into all at once, competition forces companies to begin to quantify more discrete market opportunities, and pick a point of entry into the market. In technology markets this is most commonly done through market *Segmentation.* As a result, markets diverge as the technology is applied across a potentially wide variety of disparate applications.

Segmentation marks the transition of product lifecycles into the Application quadrant, and the start of Phase II Product Lifecycles. Segmentation can be done in many ways, but often is accomplished by aligning company core competency and domain expertise with unique segment needs to create technology applications that solve specific problems. Hence, segmentation can occur through a wide variety of demographic, psychographic, geographic, and industry classifications.

As market segmentation continues and more competitors and channels divide up the market, market segments grow in number but shrink in size. This limits growth potential within each segment and forces competitors to look at alternative growth strategies.

One of the ways market expansion happens in this phase is through indirect sales and delivery channels. Channels at this stage typically hinge on specialty resellers with expertise at applying technology to meet their market's unique needs. This allows companies to scale their business into new and complementary market segments.

However, even with channels, competition still exists and many companies know that market expansion will be accomplished faster through consolidations, mergers, or acquisitions. Hence, *Consolidations* mark the next evolutionary step for product lifecycles. These consolidations enable companies to integrate both complementary and competitive products and companies into larger, more stable entities.

Consolidations have two important effects on product lifecycles: market expansion and convergence. Market expansion helps companies extend their market reach and find new applications for their products. Convergence is important because it creates product solutions that appeal to the more conservative buyers in Phase III, the Solution phase.

Products emerge into Phase III as *Product Solutions* that offer one-stop shopping for buyers. These solutions also fuel channel adoption. In Phase III, channels emerge to supplement the company's reach beyond its capacity to serve. As a result, markets can last for years in Phase III as companies periodically introduce continuous innovations or leverage new technologies to continually redefine competitive differentiation and try to prevent the market from becoming commoditized.

By the time product lifecycles reach Phase IV, the Commodity phase, industry consortiums or governing bodies have been established, or a clear market winner has emerged, to establish defined or de facto *Standards* for the technology within the marketplace. With these standards, companies are able to more readily package and deliver their solutions to commodity buyers. Ultimately, Phase IV ends with the product or solution winding up as an *Embedded* technology that spurs new product and service markets.

Product Lifecycle Case Study: Siebel Systems

Siebel Systems was founded in 1993 to address the growing need of organizations of all sizes to acquire, retain, and better serve their customers. Today, with over $1.5 billion in revenues, Siebel is a powerhouse of the Customer Relationship Management (CRM) market. According to AMR Research, Siebel has an estimated 18% of the CRM software market.

Table B highlights selected events in Siebel Systems' history as they relate to Key Market Indicators on the Product Lifecycle Curve of the Market Maturity Matrix. This table includes the date of the event or announcement, a description of the event or announcement, and the Key Indicator this event aligns with. By comparing these events with the product lifecycle curve in Diagram C, you will see the distinct parallels between Siebel's market strategy and the product market lifecycle described by the model.

Table B. Siebel Systems Events Align with Key Market Indicators

Date	Event or Announcement	Key Indicator
1993	Siebel founded; introduces Sales Force Automation product	Product Innovation
1995-1996	Siebel adds marketing automation and customer support applications	Product Innovation
1996-1997	Estimated 400 companies enter front-office automation market with a variety of Sales Force Automation, Marketing Automation, customer support, customer service, and technical support applications	Competition
November 1997	Acquires Nomadic Systems Inc., a provider of innovative business solutions to pharmaceutical sales forces; adds to Siebel's current vertical market strategy including banking, insurance, and telecommunications	Segmentation

May 1998	Completes merger with Scopus Technology, a leading provider of customer service, field service and call center software solutions	Consolidation
December 1998	Purchases The 20-20 Group, a leading provider of end-user training for the front office market	Consolidation
February 1999	Releases eBusiness applications for Customer Relationship Management	Product Solution
February 1999	Introduces SiebelNet application hosting, management, deployment, and training	Channel Maturity
October 1999	Signs agreement with IBM to jointly develop and market CRM software	Channel Maturity
November 1999	Acquires OnTarget Inc., a leading provider of consulting services and training programs for sales and marketing organizations	Product Solution
April 2000	Acquires OpenSite Technologies Inc., a market leader in Web-based dynamic commerce solutions	Product Solution
April 2000	Releases Siebel eBusiness 2000, a comprehensive family of e-business applications software	De facto Standard in CRM
February 2001	Siebel eService Wins "Best of Show" at Call Center & CRM Solutions Conference & Expo 2001	De facto Standard in CRM

In many ways, Siebel's market strategy set the pace for the entire CRM market. Siebel credits much of its success to its intimate customer relationships. To do this, as the CRM market matured, Siebel continually added customer programs to reflect changing degrees of customer involvement. Customer programs now include a user group, technology briefings, product sessions, and a technology advisory council.

Another key to Siebel's success has been its industry vertical focus, enabling it to specialize in solving problems unique to each vertical market. For instance, Siebel's banking vertical offered solutions to everything from Internet banking to telebanking, relationship banking, retail banking, wholesale banking, investment banking, check clearing, credit card processing, and all the back-end systems of the bank. Much of this was done through partnerships and alliances. Today, Siebel's estimated 600 partners and alliances include software and hardware vendors, content providers, database and operating system companies, network product and service companies, application hosting companies, strategic consulting companies, sales and service process consultants, and sales methodology vendors.

In addition, as Siebel's market matured, they continually added sales, delivery and consulting partners to complement its direct operations. Siebel partners now include IBM, Pricewaterhouse Coopers, KPMG, Deloitte Consulting, and Andersen Consulting.

SERVICE LIFECYCLE CURVE

Service Lifecycles evolve clockwise on the Market Maturity Matrix from the lower right quadrant - the Commodity quadrant - to the upper left quadrant - the Application quadrant - as illustrated in Diagram E.

Diagram E: Service Lifecycles

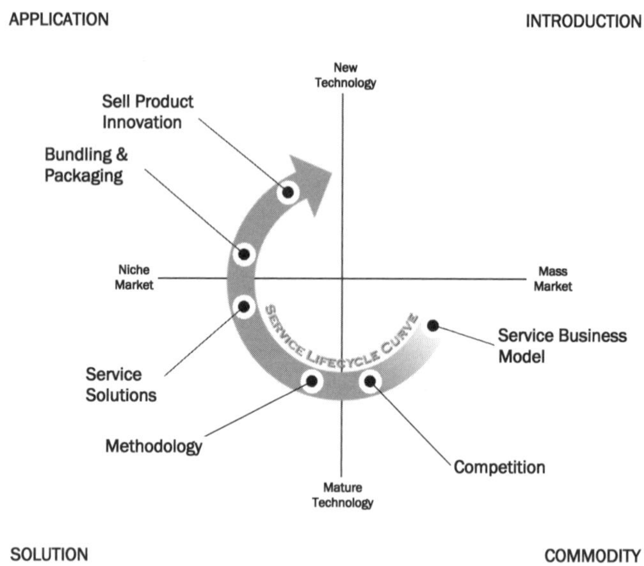

Phase I Service Lifecycles begin as *Service Business Models* based on mature or standardized technology, products, services, and channels. In fact, Service Business Models typically begin as an alternative to the product or technology they are based on such as in the case of e-procurement, outsourced web conferencing, outsourced application hosting and management, or outsourced web site development and hosting.

Since service lifecycles are born out of mature or standardized product markets, service lifecycles have few barriers to entry and tend to attract many competitors. As a result, once service providers launch into this market, *Competition* – both real and perceived – emerges quickly.

Due to this prolific competition, many service markets, without clear differentiation, suffer from poor customer loyalty, low switching costs, and rapid price deterioration. This encourages service providers to transition rapidly from Phase I to Phase II. If service markets fail to transition out of Phase I, they tend to slide into a war of attrition, where the last service provider standing – usually the one with the deepest pockets – wins.

Phase II Service Lifecycles are located in the Solutions quadrant on the Market Maturity Matrix. In this phase, where the predominant buyers are more solution oriented, vendors focus on finding ways to differentiate their services beyond price. Much like segmentation helped product marketers differentiate in Phase II Product Lifecycles, service marketers need to differentiate in Phase II Service Lifecycles. In service markets, however, differentiation is not based on segmentation, but rather on the service provider's unique approach or *Methodology* to delivering, managing, or supporting the service. Methodologies differ from market to market, but are traditionally based on a service provider's ability to package their services into " process-based offerings" that align unique competencies with specific market drivers or pains, such as service quality, reliability, or manageability. In turn, methodologies enable service providers to create competitive distinction and better identify with the needs of a portion of the market.

Once these methodologies have been created, service providers can prove their ability to consistently provide at least one service really well. This creates three important opportunities for service providers:

1. Provides a consistent and scaleable infrastructure for supporting a growing organization
2. Provides a more predictable service for the conservative buyers of Phase II Service Lifecycles
3. Makes their service more attractive to referral or reseller partners who can provide access to prospects through a trusted source

With these three elements in place, service providers can now more aggressively expand their range of services to create *Service Solutions*. Much like product solutions, Service Solutions are a convergence of complementary and competitive services that, as a whole, deliver greater value when provided by a single vendor than when delivered by discrete service providers. They offer one-stop shopping for new customers and cross-sell and up-sell opportunities for existing customers.

But service markets do not stop here. In most markets, service solutions are too easily replicated, which encourages service providers to look at ways to create a more sustainable competitive advantage. This is the opportunity of Phase III Service Lifecycles. However, transitioning to Phase III Service Lifecycles isn't easy. To do so, service providers need to transform their marketing and selling approach to align with buyers who are quite different than their previous customers. Buyers in Phase III are looking at services as a means to create competitive advantage. These buyers not only value expertise in the service, but in a service provider's understanding of the unique applications and opportunities to apply and customize this service.

This is where *Bundling and Packaging* emerges. Bundling and Packaging involve integrating specific service components along with new and emerging technologies to create service-based applications that solve specific problems or create new opportunities for their customers to build competitive advantage. This is a significant step for service providers - and one that offers tremendous potential rewards. Done properly, Bundling and Packaging create significant barriers to entry for competitors, high switching costs, and greater customer loyalty.

As a final step in service markets, services can be used to *Sell Product Innovation*. Just as mature product markets fueled new markets, so can mature service markets. Fully matured service markets create awareness and demand for new and emerging technologies and products that complement, extend, or leverage the service platform. These products can be physical products such as new hardware or software, or service products such as content as in the case of web portals.

Service Lifecycle Case Study: Amazon.com

Amazon.com was founded in 1995 with the singular goal of transforming the book buying industry. Since this time, Amazon.com has accomplished much more than this. Not only is Amazon.com the world's #1 online bookseller, it has also become the world's #1 online DVD and Video reseller. In August of 1997 it was estimated that 4.5% of all web households visited Amazon.com during that month. This represented more traffic than wal-mart.com, QVC.com, and landsend.com combined.

Amazon.com is one of the world's most recognized online brands and is the premier bookseller on most of the top Internet web sites and search engines. One of the keys to Amazon's success was its successful transition into Phase II Service Lifecycles using two key methodologies: personalized shopping and 1-click ordering™. Personalized shopping profiled a visitor's buying behavior and preferences, allowing Amazon.com to make recommendations upon the visitor's return. 1-click ordering saved time for shoppers and improved security by eliminating multiple credit card submissions over the Internet.

Amazon.com's history of market leadership is a great demonstration of a complete service lifecycle. Table C highlights selected events and announcements in the last several years for Amazon.com, showing how these events and announcements align to the Key Market Indicators on the Service Lifecycle Curve of the Market Maturity Matrix.

Table C: Amazon.com Events Align with Key Market Indicators

Date	**Event or Announcement**	**Key Indicator**
July 1995	Amazon.com introduced online book buying to market; positioned as Earth's Biggest Bookstore™	Service Business Model
1995-1996	Various book resellers launch web sites; most notably Barnes & Noble	Competition
1995-1996	Amazon focuses on personalized selling, processes for web-based payment, direct shipping, and scaleability of operations for warehousing and fulfillment	Methodology
October 1997	Announces Amazon.com is the premier bookseller on AOL.com, Yahoo!, Netscape, Excite.com, Alta Vista Search Engine, and the Prodigy Shopping Network	*Channel Expansion*
April 1998	Acquires Bookpages, Ltd. and Telebook, Inc. to expand into European market	*Channel Expansion*
April 1998	Acquires Internet Movie Database Ltd.	Preparation for Service Solution

June 1998	Launches online music store with 10 times the number of titles of a typical retail music store; offers integrated shopping cart	Service Solution
August 1998	Acquires PlanetAll (provides web-based address book, calendar and reminder service); acquires Junglee Corp. (provides web-based virtual database technology to find millions of products on Internet)	Service Solution *via one-stop-shop*
November 1998	Opens video store and holiday gift store; adds personal electronics, games, gadgets, and toys; adds gift-click to make gift giving easier	Service Solution
Feb 1999	Invests in drugstore.com to leverage customer base	Service Solution
March 1999	Invests in Pets.com to leverage customer base	Service Solution
March 1999	Opens Amazon.com Auction to help people now *sell* virtually anything online	Service Solution (and channel expansion)
April 1999	Acquires Livebid.com, the sole provider of live-event auctions on the Internet, to give local and regional auction houses full access to the Internet Auction community	Service Solution
May 1999	Introduces Star Wars store that was positioned as "entire Star Wars universe in one convenient store"; included merchandise and editorial content	Bundling and Packaging
May 1999	Expands gift store to include special sections for holidays and events such as Mother's Day, Father's Day, and Graduation Day	Bundling and Packaging
August 1999	Introduces Purchase Circles™ that creates virtual market segments. Offers top 10 music and books by geography, company, education level, government, etc.	Bundling and Packaging

October 1999	Launches Amazon.com Anywhere that provides shopping from wireless devices such as the Palm VII™ from Palm Computing	Sell New Product Innovation
February 2000	Agreement with Motorola offers bookmarked access to Amazon.com from Motorola wireless handsets and web phones	Sell New Product Innovation
March 2000	Signs agreement with Airtouch Mobile to provide access to Amazon.com via their Internet-ready phones	Sell New Product Innovation
April 2000	Announces access to Amazon.com via new Pocket PC devices such as those from Casio, Compaq, and Hewlett-Packard	Sell New Product Innovation
May 2000	Offers unique and exclusive content via Microsoft Reader eBook platform	Sell New Product Innovation
April 2001	Amazon.com and Borders.com re-launch borders.com web site as co-branded site powered by Amazon.com e-commerce platform	Product Innovation (Product Lifecycle)
May 2001	Amazon.com and babiesrus.com will re-launch babiesrus.com web site using Amazon.com e-commerce platform, warehousing facilities and fulfillment services	Product Innovation (Product Lifecycle)

Note: Items in *Italics* are not standard Key Market Indicators on the Service Lifecycle Curve, but are instrumental in market evolution.

CROSS-LIFECYCLES

One difficult question many companies face is "Are we a product company or a service company?" The answer is not always as obvious as it seems.

With the Internet, products are often marketed and sold as services using application hosting and outsourced management. Services are often packaged and delivered like products using fixed pricing and on-line ordering. There are also a growing number of markets that combine product and service components into single purchase decisions such as the wireless and satellite phone markets, the satellite television market, the Digital Subscriber Line (DSL) market, and the wireless device market such as Personal Data Assistants (PDAs) with wireless Internet.

Without the Market Maturity Matrix, cross-lifecycle markets can be difficult to interpret and predict. This makes it complicated for companies to determine which lifecycle curve they are on, how to enter the market, how the market will evolve, and what direction it is going. As these situations relate to the Market Maturity Matrix, they are cross-lifecycle markets. They show clear signs of both product and service lifecycle curves, but can move in multiple directions simultaneously.

The Market Maturity Matrix helps companies identify where the product and service characteristics of their market intersect, which direction the market is headed, and how best to apply marketing strategies and tactics. For instance, cross-lifecycle solutions rarely start in Quadrant I or Quadrant IV as new market lifecycles. Rather they begin somewhere along either the product or service lifecycle curve, automatically moved forward along the curve by the opposing lifecycle. This is somewhat of a magnetic effect between product and service lifecycles where advances in one market can have a direct impact on the other.

Cross-Lifecycle Case Study: Iridium

When Iridium entered the satellite phone service market in November 1998 they originally approached the market as an alternative to traditional cellular phone service. Unlike analog and digital wireless phones, Iridium's satellite phone service could be used anywhere in the world using orbiting satellites rather than relying on radio towers for service. For Iridium and its parent company, Motorola, the launch required an investment of more than $5 billion. This included a network of 66 satellites capable of delivering mobile phone service nearly anywhere on the globe, creating a vast distribution and reseller network, development of the satellite phones, and an expansive marketing push.

Iridium launched into the market by positioning their solution as an alternative to traditional cellular, which was being marketed horizontally to both business and consumer markets in Commodity market fashion. As a result, Iridium was forced to use comparable marketing strategies and tactics to the cellular companies including mass marketing and advertising typical in Phase I Service Lifecycles. Because Iridium had little focus to its marketing effort, it was unable to create a compelling distinction between satellite and cellular service. Worse still, this created difficulties for its channel, who had difficulty justifying to customers the $1,000 price tag for the satellite phones and the $7 per minute cost for calls.

As Iridium found out, its mass marketing approach did not result in the same results being achieved by the cellular companies. By 1999, while cellular phones had reached millions of customers, Iridium had only about 60,000 satellite service customers, which was a fraction of the business needed to provide a return on the investment made. As a result, Iridium filed for Chapter 11 and discontinued commercial service in August 1999.

Had Iridium used the Market Maturity Matrix, it may have seen that satellite phone service was not merely an alternative to the mass cellular market. It would have realized that the combination of its product and service technologies - both of which were significant continuous innovations over current technologies - were not at the same market stage as the cellular market. The market for satellite phone solutions were not in either Quadrant I or Quadrant IV where typical product and service markets begin. Their market was closer to where these two markets intersected in either Quadrant II or Quadrant III. To narrow this more, Iridium may have recognized that what they offered was not a complete solution, but rather a bundled or packaged solution that was ideal for certain market segments. In essence, it would have seen that the right market entry point for the satellite phone market was in Quadrant II, the Application quadrant.

This would have highlighted the need for Iridium to identify discrete market segments that required satellite because it offered something conventional cellular could not based on its geographic coverage and its ability to meet the unique needs of certain market segments. This information would have become the basis for quantifying potential market sizes, growth rates, and the return on investment for this venture rather than equating it to the mass cellular phone market. It would have also indicated a need to focus on channels that could add domain expertise or access to potential target market segments.

But this story looks like it may have a happy ending. In December of 2000, Iridium Satellite acquired the assets of Iridium and re-launched the service into the market in March of 2001. It did this with two significant changes to its business and marketing model. First, Iridium Satellite took a very focused approach to their marketing strategy, and identified several market segments to pursue including Heavy Construction, Defense and Military, Government, Emergency Services, Maritime (Commercial and Leisure), Mining, Forestry, Oil & Gas, Aviation (Commercial, Corporate and Leisure), Humanitarian, and Leisure & Adventure. Second, Iridium Satellite changed its distribution strategy to focus on service partners who specialized in these key target markets.

In essence, Iridium Satellite is following the exact strategies that the Market Maturity Matrix would tell them to do in the Application quadrant. As a result, Iridium is now consistently meeting its growth targets for new customers, and optimistic about its future. With its new data capabilities, Iridium Satellite is also looking to extend the applications for its satellite service into new and emerging satellite-based data services and markets.

4

Product Market Lifecycles

We were once asked to help a software company re-launch a product that was struggling, despite that it was based on a very exciting new technology. When asked for his view, the vice president of development stated to us, "We developed a great product. Sales just doesn't know how to sell it."

There was clearly a disconnect in this organization between sales and development. For whatever reason, development and sales were not jointly defining product requirements. As a result, development was creating products in expectation that sales would create a market for it.

This approach to product development is what we call "force feeding," and it rarely results in products that meet an identifiable or quantifiable market need.

The distinction this company failed to make was the distinction between "products looking for a market", and "markets looking for a product."

On the Market Maturity Matrix, product lifecycles progress through four distinct phases of evolution, moving counterclockwise starting in the upper right quadrant: Introduction, Application, Solution and Commodity, as illustrated in Diagram F.

Diagram F: Product Lifecycles by Market Phase

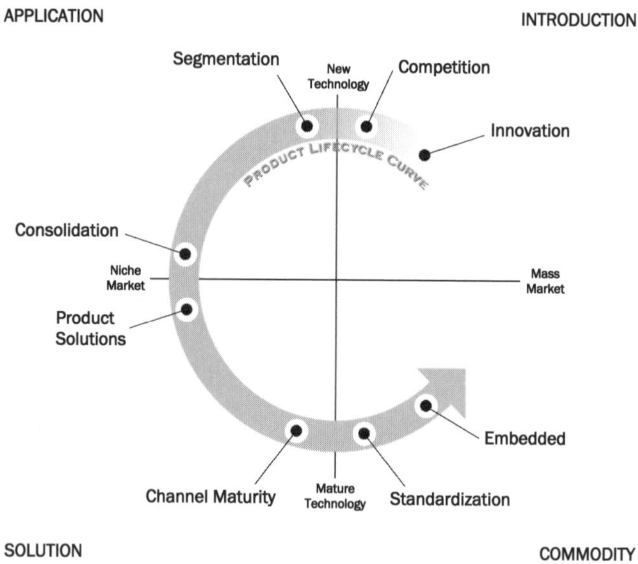

Each of these market phases offers unique windows of opportunity. For this reason, it is critical to use the Market Maturity Matrix to identify a market's maturity, predict market transitions, and apply industry best practices to seize market leadership opportunities.

PRODUCT LIFECYCLE PHASE I

One of our clients had just launched a new geological 3D modeling product . . . a true discontinuous innovation. They had received some good initial press on the product because of its highly innovative nature.

The VP of Marketing told us, "We're sure that our target customers are the Departments of Transportation at the state and federal levels. That's where we're going to focus all our efforts. DOTs were the first ones to test it and purchase it. About 75% of our customers are there."

We asked, "How many customers do you have?"

"Four," he responded.

Basing future sales and marketing efforts on the behavior of a limited number of early adopters is one of the biggest mistakes companies make. You'll find that these early adopters rarely represent the true market for your product.

Product Lifecycle Phase I is the upper right quadrant of the Market Maturity Matrix, defined by the intersection of Mass Markets and New Technology, as illustrated in Diagram G.

Diagram G: Product Lifecycle Phase I

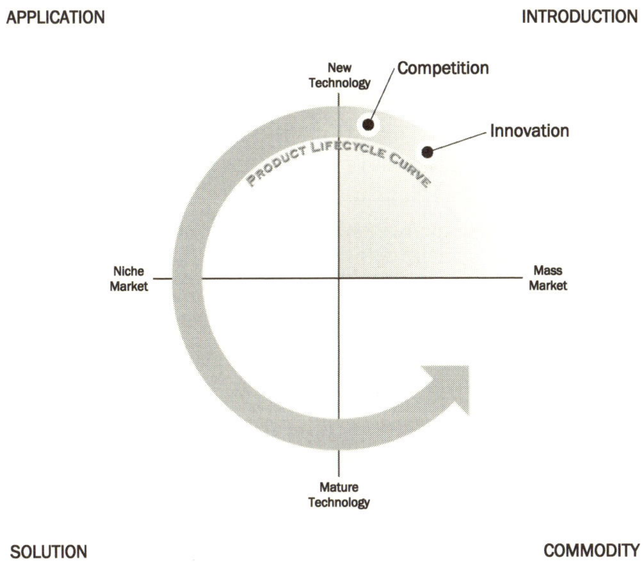

When a true discontinuous innovation appears, it appears in this phase. Using words like paradigm shift, market disruption, and market revolution, companies introduce discontinuous innovations with exuberance and optimism. They cite potential uses that cover the spectrum of business and consumer markets. For this reason when discontinuous innovations are first introduced, companies are intentionally broad in their market approach. They use a variety of inexpensive media, viral, and network marketing tactics to reach technology savvy buyers already actively seeking new innovations.

However, Product Lifecycle Phase I also poses unique challenges, which is why this phase is also a graveyard of unfulfilled technology promises, and undiscovered products and companies.

First Adopters

Technology Innovation usually is first announced through local and specialized press and media outlets, peer networks, Internet chat rooms, and "wired" sections of trade and industry publications. It generates the small articles with the big claims. The articles speak of revolutions, market disruption, futuristic applications, visionary leaders, and grandiose plans. For buyers at this stage, known as First Adopters, news of innovation is impossible to ignore. It's like a Siren's song luring in passing ships. In addition to the few, but important, enthusiasts that answer this siren song, the ones who listen hardest to these messages are potential competitors. For this reason, once the word is out, the clock is ticking.

As a result, marketers launching technologies at this stage need to be flexible, responsive, interactive, and hungry to learn from early market experience. They need to focus on activities and media that appeal to First Adopters, and appeal to their enthusiasm for new technology.

An important lesson for marketers is that First Adopters are unique in their buying behavior, and usually do not represent the characteristics of buyers at any other market phase. Hence, experience in this phase should be used to validate assumptions, not necessarily to make assumptions. For instance, early adopters, due to their passion for technology, can and will try very hard to influence product direction. To such ends, they can be very forceful and persuasive. But keep in mind that these individuals are a subset of a subset of the overall market, and their demands can often distract otherwise sound business and product plans.

Avoiding Market Temptations

Many companies at this phase enjoy initial success with their technology, either through early market acceptance or through media and analyst attention. Unqualified, this early success can be translated into unrealistic expectations for market size, growth and acceptance rates. As a result, companies often believe too much of their own press, and try to skip important market steps.

***Law of Perpetual Marketing**: "Skipping" an evolutionary market stage is rarely desired or rewarded.*

An example of this was the Apple Newton personal organizer. In hindsight, it was a product far ahead of its time. It was a Personal Data Assistant (PDA) before there was a market called PDAs. Although similar in size and functionality to many of today's popular PDAs such as those from Palm Computing, BlackBerry, and Handspring, the Newton suffered from immature technology and a syndrome we call "premature commoditization".

From a technology standpoint, at its introduction, the Newton had poor handwriting recognition software that would take several more years for the market to accept. It lacked the data exchange and synchronization capabilities with personal computers that was a key aspect of the acceptance of today's PDAs. It also had relatively poor screen resolution by today's standards, which made using the Newton distracting. In other words, it was a great concept that was pushed into commodity markets before it was ready; thus "premature commoditization".

As a result, it was a technology not ready for the scrutiny of mass markets and distribution channels, accustomed to buying mature products at the lowest possible price. If Apple had used the Market Maturity Matrix, it would have taken the Newton down a different course. It would have used its First Adopters to validate the product concept, and use this experience to identify niche markets and applications for this product. They would have then focused on creating success in these niche markets and applications to build credibility, brand, and momentum before pursuing broader markets and channels of distribution. This course of action would also have given the market time to introduce competitive offerings, while providing Apple with the time necessary to scale manufacturing and create the production economies necessary to drive down prices to a level expected by mass markets.

Keeping Your Eye on The Ball

You have probably heard the old saying, "death by success." Early success can be distracting to a business in many ways. One of the most damaging is if it focuses attention on the technology, not on the market it is intended to serve. In doing so, companies can miss the real marketing opportunity, and watch as competitors introduce products that race ahead of them into the market that they created.

***Law of Perpetual Marketing**: Technology is only an enabler. It's not a product until it serves a market.*

An example of this is the spreadsheet market. Many of you have heard the story of the first spreadsheet software, introduced by Dan Bricklin as VisiCalc in 1979 for the Apple II. VisiCalc was written up in articles as being the first application with the potential to drive demand for computer sales beyond the hobbyist and into corporate culture. It was dubbed a breakthrough for business computing. Although VisiCalc became the fundamental basis for all future spreadsheet applications, it wasn't until Lotus Development Corporation introduced Lotus 1-2-3 for DOS in 1982 that the impact of the electronic spreadsheet was truly felt. Lotus understood the applications, buyer characteristics, and delivery channels necessary to transition the spreadsheet from "gee wiz" technology into an application. In turn, Lotus figured out how to use the technology to solve real business problems. Lotus subsequently grew to $53 million in revenues by 1983, while Visicalc remained a favorite of hobbyists that eventually faded into the landscape.

Innovation

Innovation in this model refers to discontinuous innovations, rather than continuous or incremental product or technology enhancements. At this stage of the technology's maturity, companies are still not quite sure just who will ultimately use this technology or how it will be applied. Marketers of these technologies have a pretty good idea of the potential uses, but since there may not be any clearly identifiable applications for this technology yet, they have to rely on the First Adopters for help. First Adopters are the traditional visionaries, or technology enthusiasts, who are constantly looking for discontinuous innovations. These First Adopters generally have the vision to see technology not only for how it works, but also for how it can fundamentally change their existing systems, processes, and working paradigms.

Press and media, technology trade publications, technology trade shows, and industry visionaries all help generate excitement and appeal for early stage technology. As a result, First Adopters typically find you before you find them. With a good viral campaign in place, marketers at this stage can also focus efforts on potential strategic and channel partners, looking for, among other things, funding and validation.

Competition

At the end of Product Phase I, competition - both real and perceived - emerges as companies race to announce their intentions. Part of the reason for this rapid competitive pressure is that very few technologies can remain unique or proprietary very long. Despite patents, trademarks, and intellectual property rights, nearly any technology can be duplicated or reverse-engineered. Few discontinuous technologies can enjoy success the way Novell did in the early 1990s and Cisco did in the late 1990s. The reality is that technology leadership is fleeting, and can usually be counted in months, if not weeks.

Depending on the barriers to enter a market, product markets can get crowded pretty fast. However, competition isn't bad. Competition is necessary to fuel a market. Hence, companies not only need to welcome competitions, but also embrace them.

***Law of Perpetual Marketing**: Competition is necessary and should be embraced.*

One important change resulting from competition is pricing. Pricing is often unique at this stage of product lifecycles, and is typically not a good indicator of pricing in future phases. In fact, until competition enters a market, the price for the technology itself in this phase is often higher than at any other point in a product's lifecycle. This is due, in part, to the fact that purchases at this stage of product lifecycles are generally more emotionally driven. This decreases price elasticity and allows marketers to set pricing on a more value-based pricing model. However, once competition enters the market, buyers have a frame of reference for price comparison.

As an example, the first microwave oven was introduced in 1952, marketed as the Tappan Range. It was priced at $1,295, which, particularly in those days, was a lot of money. Today, because the microwave oven market, its distribution channels, and the uses for this technology have matured to a state of commodity, prices for microwaves begins at under $100. Similarly, flat panel monitors for desktop computers introduced in 1999 were priced starting at $1500-$2000 for a 14" diagonal screen. Within eighteen months, as flat panel monitor sales matured out of phase I, prices dropped steadily to less than a third of their original price.

There are certain red flags that we look for when we talk with technology companies. One of the red flags we often look for in this market phase is when a CEO proudly says, "we have no competition." Our typical response to this is, "well, you better find some." You shouldn't fear competition. Learn to embrace it and leverage it. Competition is necessary for creating a market and the momentum that helps them mature. Remember, the more unique your product, the more unique your customer. In other words, if you are the only company doing what you are doing, there may be a good reason. Without competition, you're not in *a* market – you're just in *the* market. You need more than one company to make *a* market. So, if you aren't sure with whom you are competing – find out.

PRODUCT LIFECYCLE PHASE II

A client of ours had created a truly unique tool for web development. After a year, the market for this product remained in Phase I of the Product Lifecycle. Despite this, the company was committed to launching a new version that would give this advanced multimedia development power to virtually anyone.

The CEO told us, "This product will be so innovative and easy to use that absolutely everyone will want it."

As we educated the CEO and his team on the Market Maturity Matrix, it became clear that although the product would be easy enough for just about anyone to use, the reality was that the market was just not ready yet for these capabilities. What was exciting, however, was that there were various market segments that would pay top-dollar for this product because it would give them a competitive advantage.

With this insight, the CEO was the first among his team to identify that the market was clearly in Phase II and immediately understood why companies in this phase do what they do.

Product Lifecycle Phase II is the upper left quadrant of the Market Maturity Matrix, defined by the intersection of Niche Markets and New Technology, as illustrated in Diagram H.

Diagram H: Product Lifecycles Phase II

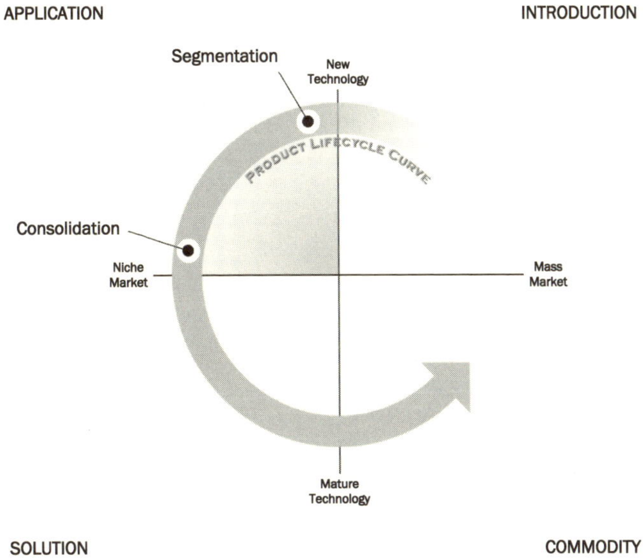

Inertia is a term used to describe the momentum of an object in motion. The faster and heavier the object, the greater its inertia. Inertia for technology markets works in a similar fashion. When product lifecycles transition into Phase II of the Market Maturity Matrix, they need a certain amount of inertia to sustain market growth and user acceptance. The more companies competing in the market, the more coverage the market and technology receive from analysts and media, the greater its inertia.

In addition, individual companies competing in this market phase also need inertia. Transitioning from Phase I to Phase II requires momentum to make changes in business models, sales strategies, operations, and financing, among others to meet the changing technology, application, and business needs of buyers in this market phase.

Phase II is a company's opportunity to transition its market development strategy from being a "mile-wide and inch-deep," into an "inch-wide and a mile-deep." This is the essence of Segmentation, the first of the two market indicators in Phase II. However, as market segments mature and the growth potential within each segment begins to shrink, markets go through a period of convergence as channels and buyers mature and vendors look at new ways to expand their product and their market share. One of the most predictable ways this happens in technology markets is ırough Consolidations. Consolidations refer to product as well as company convergence, including product integration, mergers, and acquisitions.

Technology Movers

As product markets transition into Phase II, buyer behaviors change. The First Adopters that dominated Quadrant I are replaced by the Technology Movers of Quadrant II. Technology Movers are not looking for pie-in-the-sky, first-of-its-kind technology innovation. They are looking for technology and application components - products that complement existing initiatives to act as change agents for their business.

Technology Movers are primarily driven by a desire to create competitive advantage in their marketplace. To do this, these buyers are willing to accept a certain degree of risk in implementing these products; and in return, expect a high degree of domain or application expertise in their industry or business function from their vendor to mitigate this risk. This, in turn, forces markets to diverge and segment based on unique applications and uses of the technology. This drives companies to focus and target market segments, and build expertise in these areas.

For example, in Phase I, the market for CD-R (recordable CD-ROM drive) technology was full of potential and First Adopters were the sparks that brought the market to life. In Phase II, CD-R marketers targeted segments of Technology Movers with specific applications that would create value for this product in specific applications. For business markets, this included faster duplication of production software, rapid creation of sales tools such as demonstrations, and faster data backups. For consumer markets, it included mixing personal music CDs and home photo albums. Different manufacturers tailored their product names, pricing, packaging, positioning, messaging, and channel distribution based on these target segments.

Making Tough Decisions

It is fair to say that transitioning from Phase I into Phase II often forces companies to make tough decisions about how to select and prioritize markets, as well as how to build the knowledge necessary to align with each market segment's unique needs. By properly segmenting the market and selecting markets based on alignment with core business goals, values, and competencies, successfully maneuvering into Phase II can be significantly less daunting and much more predictable.

As companies move into Phase II, more structured processes and practices become more important. Looking back at Phase I, companies could

afford to work without rigid rules or processes. In fact, in Phas
works to a company's advantage because they need to rapidly learn from experience, and turn this into product and technology changes. But in Phase II, processes help to create repeatability and scaleability. For marketing this includes tracking and measuring results more rapidly to understand changes in the market and its buyers. For operations this includes processes in key areas such as product development, order processing, account service, product delivery, and technical support.

Segmentation

One of the keys to transitioning into the Application Phase is targeting segments of the market with unique value propositions. Segmenting markets is essential to successfully identify early market applications and target discrete audiences with a value proposition that is unique and defensible from the competitors. For many companies, segmentation is an ugly and intimidating word. It means artificial limitations and unnecessary barriers. Sales organizations and executives often feel limited when confined to targeting specific markets. However, marketers know that segmentation is vital for success. Although many businesses want to remain open and flexible to business outside of specific market segments, marketing will thrive in an environment where it can target discrete customer sets. For marketers, segmentation is a constant companion, and an important weapon in the arsenal for focusing and prioritizing efforts, attracting the right prospects, and identifying with customer needs. Segmentation is also the key to creating early product or company messaging and positioning.

Some markets at this stage segment themselves - usually by application need, relative pain, or urgency. For markets that don't segment themselves, it is up to the marketers to define logical segments, taking into consideration demographic, psychographic, and financial market dynamics. This is where the tough decisions get made. By packaging products as applications for solving specific market needs, marketers can more precisely define target audiences, motivations, business pains, and the appropriate tactics for speaking to these individuals. This includes capitalizing on existing or attainable domain expertise, as well as an understanding of the customer applications for the product. The most successful companies begin with segments that best match their expertise or competitive advantage.

The caveat with segmentation is that it has to strike a balance. Segments have to be big enough to justify the expense of targeting them; but they need to be small enough to identify with a defined target market. Through segmentation, companies know where to build product, implementation, industry, or application expertise to create the domain knowledge desired by customers. By limiting the number of target markets, companies minimize the chance of diluting their domain expertise and value proposition, and minimize the risk of attracting excessive business than can be unprofitable or distracting.

***Law of Perpetual Marketing**: Segmenting and targeting markets is difficult but necessary.*

A rule of thumb for technology companies is "the larger the market appears, the more important it is to focus." Here is the segmentation process in a nutshell. First, using secondary market research, break down the market into smaller, logical market sets based on identifiable geographic and/or demographic characteristics. Second, where possible, use primary research to further qualify these segments by psychographic characteristics such as risk tendencies, financial aggressiveness, pains and drivers, sales and delivery channels, motivations, urgency, functional requirements, and key buyers and influencers. Third, group and rank these segments based on the comparison of demographic and psychographic similarities. This process helps to identify opportunities to leverage success and experience across subsequent segments by replicating your business model, approach, and value proposition.

Many companies fail to transition successfully into Phase II because they fall victim to the allure that "one size fits all." In other words, they take a horizontal approach that inevitably has to accommodate the lowest common denominator across segments. The result is often mediocre messaging, poor differentiation, and weak identification with a target audience. We call it the "something for everybody, everything for nobody" syndrome.

Consolidation

Throughout Phase II, each competitor in a market has its respective market segments that it is aggressively trying to grow through its domain expertise, application expertise, service offerings, and channels. However, as each of these market segments matures, the opportunity for growth within each segment begins to shrink. This causes companies to more aggressively expand outside their core market segments, creating increased competition.. With more competitors to consider, buyers begin to put pressure on competitors for more buying convenience and clearer competitive choices.

In response, market convergence begins lead by consolidations such as mergers and acquisitions.

It is often quoted that 75% of all mergers and acquisitions fail. But in technology markets, where failure rates despite consolidations are equally intimidating, consolidations happen with fierce regularity. Despite this, consolidations such as strategic and tactical mergers and acquisitions, as well as strategic partnerships and alliances are a predictable and healthy stage in a market's evolution. Examples of consolidations include: the acquisition of a complementary company; the integration of two or more complementary products; the acquisition of a Value Added Reseller (VAR)in a specific market segment; the acquisition of a consulting organization with specific domain expertise; or a reseller agreement with systems integrator.

Once consolidations begin in a market, subsequent ones usually happen in rapid succession. Companies who once believed they could grow fast enough organically suddenly realize they cannot. Markets begin to resemble a race for dance partners before the last song of the night. The companies that typically win this race are the ones with the deepest pockets (or currency such as stock), the most strategic partners and alliances (range of merger or acquisition choices), the most pervasive channels, or clear product superiority. Thus, cash reserves and strong alliances with complementary product and service providers are an essential part of Phase III market planning.

PRODUCT LIFECYCLE PHASE III

We had a client that provided software for Internet-based meetings and presentations. The CEO was discouraged because over the past several months the company was not able to charge what it used to for its product. "We're finding that new customers are unwilling to pay what existing customers did for the same product just six months ago. And, existing customers are demanding enhancements faster than we're able to develop them."

"That sounds about right," we said.

"Right?" the CEO said surprised. "It's very wrong as far as I'm concerned."

"But, you have to understand what's happened to your market," we said. "There are two forces at work here. First, the people who purchased your products six months ago were

> *looking for a competitive advantage and were willing to pay a premium for it. New customers are simply looking for competitive parity. In other words, they're seeing everyone else in their market using technology similar to yours, and they're trying to catch up. At this point in the market maturity, they're not going to pay that premium price. And second, it's likely your existing customers are now looking for more complete solutions. And, if you can't converge your products with the right complementary components into solutions fast enough to meet the market demands, there are few barriers to prevent them from going someplace else to get it."*

Product Lifecycle Phase III is the lower left quadrant of the Market Maturity Matrix, defined by the intersection of Niche Markets and Mature Technology, as illustrated in Diagram J.

Diagram J: Product Lifecycle Phase III

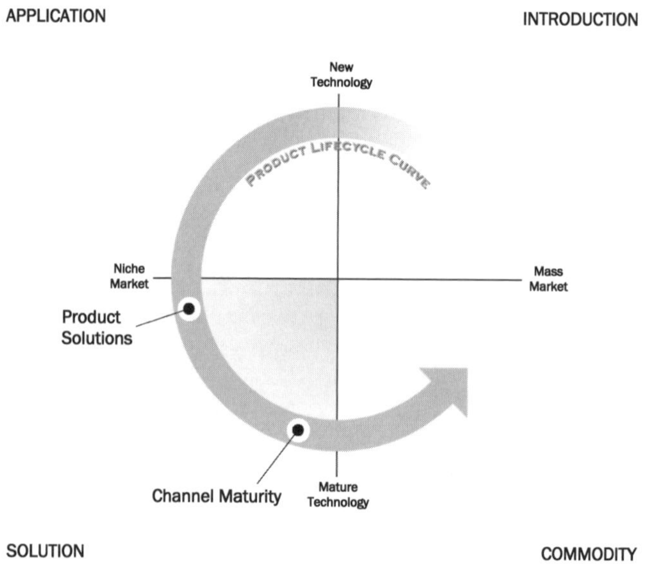

At certain times in market lifecycles, products, services, and technologies reach a point of convergence; a point at which target audiences no longer place value on the individual components as much as they do the sum of the parts.

In technology markets, this is exactly what happens as product markets transition from Phase II to Phase III on the Market Maturity Matrix. At this stage of market maturity, markets are concurrently transitioning from niche to mass market momentum, buyer characteristics are changing to reflect more conservative purchasing behaviors, channels are maturing, and the competitive landscape is stabilizing. As a result, demands changes from component value to solution value.

By predicting these changing business needs and building on a solid infrastructure of processes and practices, companies can speed through this market convergence and address the two major market changes in Phase III, the emergence of complete Product Solutions and Channel Maturity.

Market Movers

Phase III buyers, called Market Movers, take a more conservative, operational approach to technology purchases. No longer are purchasers motivated by the value of the discrete components of a solution; they are motivated by the simplicity, synergy, and accountability of a single integrated solution from a single vendor.

Market Movers, like Technology Movers, buy for strategic reasons. However, Market Movers are motivated by different drivers and behaviors than Technology Movers. For instance, Market Movers are more interested in market trends, rather than technology trends. They prefer total solutions rather than solution components. They often purchase for competitive parity reasons - they do not want to get too far behind their competitors in technology adoption. Market Movers are more financially risk averse than Technology Movers. They are more likely to use product solutions to automate or improve existing processes and practices rather than as change agents. Also, since these buyers want more complete solutions, they also tend to look to integrators or service providers to be involved in evaluating and delivering these solutions.

The Significance of Convergence

In technology markets, company consolidations, mergers, acquisitions, and fall-out happen regularly. However, as markets transition to Phase III, these events take on a different significance. Hence, they are often considered part of the market's convergence.

Convergence is important for two reasons. First, convergence allows companies to expand selectively beyond discrete segments - which was important for maintaining inertia in Phase II. Second, convergence is the technology market's equivalent of Charles Darwin's theory of natural selection. Convergence distills a wide range of applications, technologies, and related integration and implementation services down into a couple dominant solution options for the Market Movers.

As markets transition into Phase III, market expansion takes priority as competitors begin to race to minimize the impact of price reductions, leverage economies of scale, and secure channels. The strongest competitors offer complete Product Solutions at a scale and at a volume that drive growth and market expansion. In most markets, only a couple companies will survive this market convergence process to become market leaders and reach the Channel Maturity necessary to expand the market concurrently across new geographies and new industries.

As a result, Market Movers in Phase III have fewer, but more robust competitive choices. Competitive differentiation among these remaining competitors heats up, and branding begins to play a more important role in market and channel development. Smaller competitors that are unable to compete at the scale of market leaders tend to return to Phase II strategies, such as focusing on select niche markets where they can provide differentiation and high value segment-based applications, domain and expertise.

Product Solutions

Product Solutions are in many ways a direct response to the conservative nature of Market Movers looking for consistent and predictable results from technology purchases. Product Solutions simplify purchase decisions, reduce integration challenges, and minimize implementation risks by combining complementary applications with the customization and consulting services, delivery methodologies, education, and training.

***Law of Perpetual Marketing**: Product Solutions are the convergence of complementary applications and professional services into a single, integrated purchase decision for buyers.*

For instance, when Sales Force Automation (SFA) converged with Enterprise Support and Marketing Automation applications, a new category of solution was created called Customer Relationship Management (CRM). This transition was an important time for vendors such as Siebel Systems, Onyx, and Pivotal. They used this market transition as a time to educate and influence analysts and media which helped them emerge as thought leaders in their market. In doing so, they left behind many smaller, independent

SFA, customer service, and marketing automation companies who were unable to translate this market transition into a market opportunity.

Channel Maturity

As markets mature and expand, companies look to capitalize on this growth by reaching new geographic regions, new market segments, and new sets of buyers. To foster this growth, companies turn to the growing and maturing indirect sales channels in this market phase. These channels include retail stores, Value Added Resellers (VARs), Systems Integrators (SIs), Original Equipment Manufacturers (OEMs), Application Service Providers (ASPs), and more.

Based on the maturity of the technology and the sophistication of the applications or products, different channels will adopt products at different times. Each successive channel will offer a greater and greater reach into the market. Back in Phase II, specialty or verticalized resellers were generally the first to adopt new products because their business models are based on their expertise in certain markets. However, in Phase III, as products mature and grow in popularity and stability, distributors, regional resellers, service providers, and systems integrators jump on board and look to find ways to complement these solutions by combining them with other products and services already in their arsenal. By leveraging their existing customer relationships, these channel partners are instrumental in reaching customers accustomed to buying through their trusted vendor. Thus, by the end of Phase III, channel maturity has helped the market reach a new level of popularity, awareness and acceptance.

PRODUCT LIFECYCLE PHASE IV

We worked briefly with a company that offered Internet-enhanced teleconferencing. The product component allowed you to view over the Internet who was on the conference call. It also allowed you to manage each caller's ability to participate and vote on topics being discussed. Although the software worked over the Internet, it did not run in a browser; it had a proprietary interface.

The CEO was discouraged because he was unable to raise additional funding to build infrastructure to market the product.

"The problem," we told him, "is that the teleconferencing market is at a pure commodity stage. At this point in the

market's maturity, it's unfortunately a game of price. In other words, who can offer teleconferencing at a penny less per minute per connection than the next guy."

The CEO agreed, but said, "But our interface is so innovative. Once people use it, we're sure they'll love it and be willing to pay a premium for it."

"Unfortunately," we said, "market forces are contrary to that. Although your interface is innovative, it is non-standard. When people hear Internet-enhanced teleconferencing, they expect that the product will run in standard browsers. Introducing a non-standard into a market that has already standardized usually is a losing battle."

Product Lifecycle Phase IV is the lower right quadrant of the Market Maturity Matrix, defined by the intersection of Mass Markets and Mature Technology, as illustrated in Diagram K.

Diagram K: Product Lifecycle Phase IV

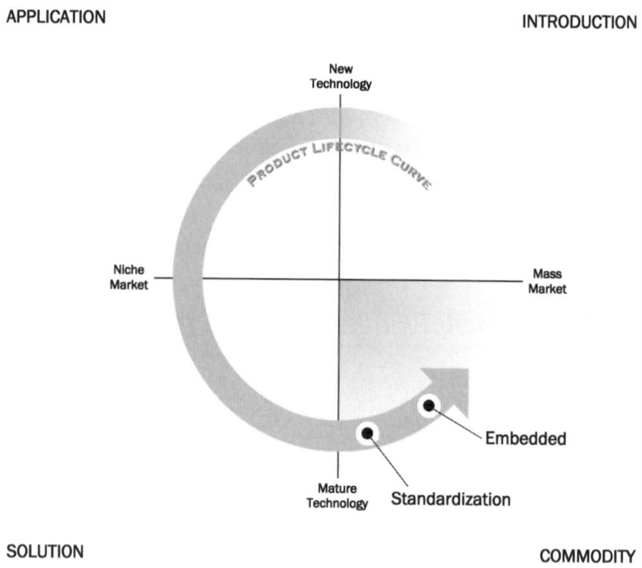

In technology markets, nearly every product has either become, or is on its way to becoming commoditized. Although many companies try to stall this, they can rarely prevent it. Commodity products, based on standards, are an important part of the incubation process for new products and services that leverage standards as components or embedded technology.

In reality, most of the technology-based products you use in your daily lives such as scanners, printers, monitors, word processing software, pagers, and e-mail packages are in commodity markets. This is, in part, due to technology Standards, or Embedded technologies, which are key indicators of the product lifecycle Phase IV.

As markets approach the end of Phase III, they are often dominated by a select group of market leaders. Some vendors and analysts argue that this indicates the beginning of the end for a market. But for many companies, it is only the beginning. In fact, many Commodity markets are very large, active and lucrative markets.

Late Adopters

At the start of Phase IV product lifecycles a new type of buyer emerges called Late Adopters. These buyers are the diehard conservatives. They tend to be motivated more by tactical gains than by strategic gains. They typically view technology more as an economic opportunity than as a strategic opportunity. For this reason, Late Adopters will wait to adopt technology products until it becomes a necessity, until it becomes more cost-effective than continuing their current method, or until their current solution is no longer supported. Many will even wait to purchase until the technology is fully embedded in another product. In turn, this helps these Late Adopters minimize risks associated with either the technology or their choice of vendor.

Late Adopters also place value on brand recognition - for both the vendor and the channel through which they purchase. Because of this, even lesser known brands in many categories can enjoy success in the Commodity Phase because they benefit from combination of the standard they support together with the brand and credibility of the channel they sell through. For instance, a buyer may place a perceived association of quality for any product bought through Software Spectrum. So if Software Spectrum carries a particular product, that product enjoys the advantages and disadvantages of this reseller. The same goes for VARs, system integrators, strategic partners, and application hosting providers.

***Law of Perpetual Marketing**: For commodity markets, brand is king.*

The End of the Road?

One of the other interesting characteristics of Late Adopters is their demand for more parity between remaining competitors. As a result, markets respond with Standards.

With standardization, whether a defined or a de facto standard, products and solutions face unique challenges. As markets become more standardized, they become commoditized, bringing with it lower margins, and less distinctive differentiation. In turn, these commodity markets tend to favor companies who can establish and maintain an operational or cost advantage, those with the most pervasive and loyal channels and customers, and those with the strongest brands.

In many cases, standardized technologies become Embedded into new, emerging products and services. With Embedded technologies, new product and service markets begin to take shape that leverage the market's acceptance and growing reliance on this product or solution.

Standardization

Standards can emerge in various ways. Two of the most common ways are defined standards and de facto standards.

Defined Standards

Defined standards are typically created by an authorized group of market leaders and representatives chartered to define a single set of technology guidelines. This, in turn, simplifies the adoption and use of products and technologies by Late Adopters. Standards can range from user interface, to connectivity, to transfer speeds, to nomenclature; just about anything. TWAIN is an example of a defined standard for scanning and faxing. So are SCSI and Universal Serial Bus (USB) for peripheral devices. TCP/IP is a standard for network packet handling and transport. Hypertext Marlup Language (HTML) is a defined standard for web site creation. The .com, .net and .org top-level domain designations are standards for Internet domain names.

De facto Standards

De facto standards are the standards created by default to conform to a dominant market leader. WinZip is a good example of a de facto standard for file compression. The MP3 file format is a de facto standard for audio file storage and playback.

Standards are an important step in technology evolution. Standards enable mass-market buyers to purchase with more confidence and predictability through their preferred channels. Standardization also helps to drive down prices for purchasers.

The flip side of this for marketers is that creating product differentiation is often difficult at this stage. Differentiation usually comes down to a "parity plus one" scenario, where the "plus one" is a single point of differentiation that although, by itself, is not a motivator to purchase within the category, it can play a role in the final decision stage of the buying process.

Embedded Technology

Once a technology has been standardized, many products spend their golden years as embedded components inside other commoditized products or solutions. For instance, scanner technology gets built into printers; CD-ROM drives get added as standard equipment to computers; and spell-checking software gets embedded into word processors. This eliminates the decision criteria a buyer would have to go through to acquire these technologies separately. Many commodity phase buyers will wait until technology becomes embedded before they place any value on it.

For marketers, embedded technology is an opportunity to establish a new annuity business. By keeping a close eye on emerging products and markets, smart marketers can significantly extend the life of commodity products through OEM, licensing, and other alternative channels where technology components are essential building blocks for new products and services, and great "parity plus one" differentiators for other Phase IV products.

Another example of Embedded technology is cellular phones. Not many people stop to appreciate the lifecycles of all the components inside their cell phones, including the microprocessors, circuit boards, Liquid Crystal Display (LCD) screen, numeric touch pads, antennas and more. Similarly with digital cameras, few buyers appreciate the distinction in quarter panel displays, storage media, or the standards in both hardware and software that were established to transfer digital images to computers; still each of these components competed in its own marketplace to reach the embedded phase.

5

Service Market Lifecycles

We once worked with a start-up telecommunications company that provided voice, data, and broadband services. They had a central marketing team representing all of the services, and to reach economies with their efforts, this team used many of the same marketing tactics and programs for all of the services.

Their programs were meeting with mixed results. Some programs were working better with some services, other programs with other services, and they couldn't explain why.

Using the Market Maturity Matrix, we helped them locate the market maturity for each of the service markets, and to their surprise, the markets were in distinctly different market phases. This helped them identify what marketing activities should be emphasized for each service. In turn, this helped them realize why some of their activities were working better than other activities, and where their priorities for each service needed to be going forward.

Like technology product lifecycles, service lifecycles also follow predictable evolutionary paths. This includes nearly all technology-based services such as application outsourcing, discrete and recurring operational services, resource-based services, and Internet and communications services.

However, service lifecycles follow a very different evolutionary path from product lifecycles. In fact, on the Market Maturity Matrix, service lifecycles mature in the reverse motion of product lifecycles, as illustrated in Diagram L.

Diagram L: Service Lifecycles by Market Phase

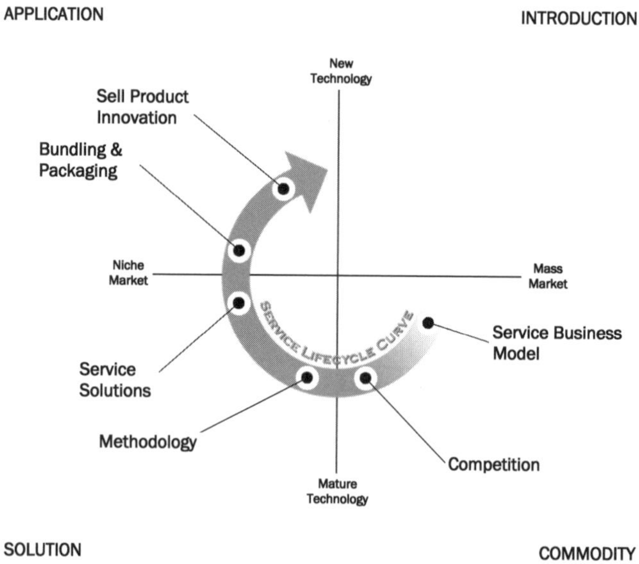

For service lifecycles, there are three distinct phases of evolution: Commodity, Solution, and Application. Beginning in Quadrant IV, service lifecycles follow a clockwise path through the Market Maturity Matrix, moving from Quadrant IV into Quadrant III, then into Quadrant II. Each one of these phases is important and necessary for the natural progression and evolutionary growth of service markets.

SERVICE LIFECYCLE PHASE I

We worked with an application software company whose products were very powerful, but also very complex. The company competed in a very mature market, but was having trouble with customer satisfaction because the implementation and ongoing management of the product was complicated and time intensive. The customers' IT departments didn't have the human resources nor the expertise to effectively implement and manage the product, and were opting towards less complicated alternatives, including Application Hosting. Even the company's consultants, who specialized in the applications, were unqualified to manage these implementations.

As a result, the company was sending developers, QA staff, and support people to customer sites in an ongoing effort to stabilize, reconfigure, and maintain installs. This was not only frustrating for managers whose resources were being used, but it was creating too many unprofitable customers.

"By its very nature," we said, "The product is going to be complicated to install and manage. But, who knows it better than your own developers, QA people, and support people?"

"Well, that's to be expected. But those resources aren't billable," he said.

"Then, why don't you take that expertise, create a professional services group, and sell those services. You can create an outsourced installation and management service group whereby your customers will get their applications up and running faster, with access to dedicated skilled resources to maintain the implementation, either remotely or on-site. This should result in better customer experiences, faster time to results, and better reference accounts. In turn, it will concurrently provide you with both new revenue streams and better customer satisfaction. As an added bonus, it will also increase the switching costs and barriers to entry by offering something of unique value."

The CEO thought for a minute, looked at his team of executives, and said, "As of today, we're in the service business."

Service Lifecycle Phase I is the lower right quadrant of the Market Maturity Matrix, defined by the intersection of Mass Markets and Mature Technology, as illustrated in Diagram M.

Diagram M: Phase I Service Lifecycle

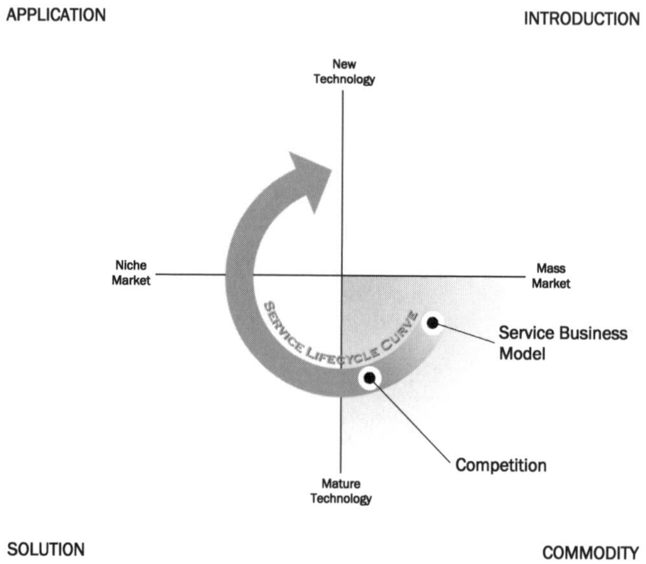

Service lifecycles start where most products end up – in mature markets. This is because technology-based services tend to be born out of mature or commoditized product markets. As a result, service market lifecycles begin in the lower right quadrant of the Market Maturity Matrix – the Commodity quadrant.

By leveraging mature products and technologies, many service markets begin as outsourced products or solutions, or as services based on commoditized products or technologies. In essence, a Service Business Model. In doing so, service vendors leverage the Commodity market's acceptance and understanding of mature technologies, while offering economies of scale, more rapid deployments, and expertise in this field. In the traditional Service Business Model, services are offered for a recurring fee. However, with the Internet, the way services are procured, priced and delivered can vary. Examples of Service Business Models includes discrete and recurring services such as:

- Application outsourcing (i.e., procurement, customer relationship management, etc.)
- Managed application, equipment and communications outsourcing (Application Service Providers, Managed Hosting Providers, Storage Hosting Providers, Co-location providers, etc.)

- Service outsourcing (i.e., call center, IT services, etc.)
- Internet services (i.e., design, management, hosting, etc.)
- Internet and broadband subscriptions
- Wireless services

Late Adopters

The predominant buyers in Phase I Service Lifecycles are driven by the financial, tactical, or short-term gains inherent in service outsourcing. This includes reduced initial investments, reduced internal resource requirements, and faster implementations. This behavior is consistent with the buyers at the Commodity phase of product lifecycles. These buyers view outsourcing services or pay-as-you-go models as a way to minimize the risks associated with implementing and maintaining new solutions.

For this reason, it is important at this market phase to appeal to the needs of traditional Late Adopters, who are financially driven and very conservative. This includes appealing to them with a proposition of reduced risk, faster time-to-market, minimized operational impact, better economies of scale, application expertise, and financial advantage. Financial buyers are usually key influencers or decision-makers in this purchase and in many cases are not looking to use services as a core or strategic part of their operations. Late Adopters are typically more focused on tactical, short-term gains.

Service Business Model

The appearance of service business models is the first step in service lifecycles. Examples of technology-based service markets include:

- Application Service Providers such as Corio and US Internetworking
- Internet Service Providers such as AOL, MSN, and Earthlink
- Managed Hosting Providers such as Exodus, Digex, and Genuity
- Co-location and hosting companies such as Sprint and Level 3
- Application Management Providers such as Loudcloud
- Application and web development companies such as Razorfish
- Business service providers who offer outsourced IT and business processes such as payroll, expense management, and HR
- Outsourced resource and IT staffing providers such as Teletech

Service markets typically emerge from product markets that have reached maturity or standardization. This is particularly true for solutions with long implementation cycles and high acquisition costs, as well as those requiring high degrees of expertise or large numbers of dedicated resources.

An example of this is e-procurement. E-procurement, part of a Business-to-Business (B2B) commerce solution promises to reduce procurement resources and costs, while improving efficiency. This is a very compelling argument for both direct and indirect purchases. However, even with an e-procurement platform such as Ariba's, its implementation and management is inherently complex. As a result, companies like Ariba have enlisted hosting or ASP partners who can not only minimize up-front IT complexity and customer investment, but extend the solution with procurement expertise, access to suppliers, and greater economies of scale.

Outsourced service business models emerge more rapidly in some markets than others. One market that matured rapidly into a service model was outsourced web development. Part of this was due to the rapidly changing tools, technologies, platforms, and standards in this space. Also contributing to this was the limitation of skilled resources, which made internal staffing of this function difficult and expensive. Hence, early in this market's introduction, a company's inability to keep pace with these changes was enough to justify the outsourcing of this service.

Competition

Service markets attract a lot of competition – partly because service business models promise strong recurring revenue streams, and partly because there are few barriers to entry. Since service markets are born out of mature or standardized products and technologies, the cost to enter service markets is usually quite low. Market entry costs for service markets typically include 1) Domain Knowledge (knowledge of the underlying technology and applications); and 2) Access to the technology source (relationships with those companies who provide the underlying technology, platform, product, or service footprint).

***Law of Perpetual Marketing**: Service markets invite many competitors due to few barriers to entry and attractive recurring revenue streams.*

Competition in this market typically includes a mix of product and technology companies, Value Added Resellers (VARs), larger service providers looking to extend their current range of services, and spin-offs of internal departments (service departments, IT departments, etc.) that are not core to the parent company. Still others appear as start-ups – teams of industry or technology experts who can add specific service value to a base technology, product or solution.

The greatest competitive threat facing service providers is that many service markets have little or no differentiation in terms of technology or market reach. As a result, price becomes the primary differentiator. This diminishes a service provider's ability to create a unique and defensible position that emphasizes and leverages a core company competency. By focusing on price, service business models rapidly lose their attractive revenues, businesses are too easily copied, and brands are too easily diluted.

Succeeding When Others Fail

Success in most service markets is an act of sheer will. It takes persistence, determination, courage, good strategy, and a bit of luck. This includes a unique and defensible market position and brand, repeatable and scaleable processes, proprietary access to unique technology, and domain expertise unavailable anywhere else. Without one or more of these advantages, most service providers will not survive beyond Phase I Service Lifecycles.

However, in some service markets, winners are determined by default. In these markets, it is a war of attrition, and the company that wins is the one that has one or more of the three key advantages to service market success: 1) more money, 2) an existing customer base, and 3) access to an established sales and delivery channel.

An example of this is the Digital Subscriber Line (DSL) market. When this market was first established, DSL service providers emerged in waves. It seemed everyone was simultaneously in the DSL business. This was because there were few barriers to entry to this market; virtually anyone could resell DSL service. The downside, of course, was that most DSL service providers had the same constraints in terms of footprint, bandwidth, and service source. For this reason, price differentiation and "parity plus one" techniques like free modems and free installation seemed to dominate this early market because there was little or no differentiation in service.

For most early service providers, the goal was to secure as many subscribers early on as possible, at any cost. Essentially it was a "land grab". But since early adopters in service markets are more tactical in their purchase behavior, with little loyalty to the service provider, margins eroded quickly as customer turnover rates and acquisition costs soared. In turn, those providers unable to make the transition into the next market stage shut their doors, unable to turn a profit from their model. Even seemingly well-funded companies ran out of money trying to 'buy' market share based on a future revenue model. In essence, this market became a war of attrition, and the last business standing was going to win. In the case of Digital Subscriber Line (DSL) service, this would be the one with the deepest pockets or the largest installed base. In Colorado, within a period of six months between late 2000 and early 2001, four major DSL providers went out of business, including Jato, Rhythms, e-Soft, and AtLink. Collectively, this represented about $1 billion dollars in venture capital, gone.

SERVICE LIFECYCLE PHASE II

When Jeff and I formed Maccel Group, we were basically marketing consultants – which are a dime a dozen. Every one of them, including us, could provide good results for clients. But because consulting was primarily based on an individual's experience, there was no way to ensure consistency or predictability of results from client to client.

With the Market Maturity Matrix, we changed all that. The Market Maturity Matrix allowed us to differentiate ourselves because we had a methodology that was reliable, repeatable and predictable. It offered a unique scientific approach to help our clients do "the right marketing at the right time." Combined with our other methodologies for branding, product launches, tactical planning, company and product naming, etc., we could achieve more predictable results, faster.

In short, our methodology was our differentiator and our competitive advantage. It allowed us to progress our market from Service Lifecycle Phase I to Phase II, and create differentiation between "marketing consulting" and "marketing acceleration," while enabling us to target whole new markets and channels with a unique value proposition.

Service Lifecycle Phase II is the lower left quadrant of the Market Maturity Matrix, defined by the intersection of Niche Markets and Mature Technology, as illustrated in Diagram N.

Diagram N: Phase II Service Lifecycle

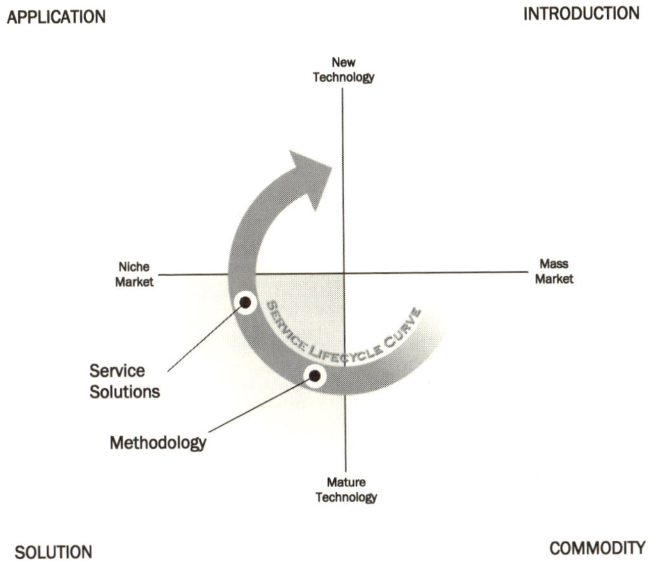

To enter Phase II Service Lifecycles, service providers have to undergo a significant change in how they package, position, and sell their services. Due to the pervasive competition in early stage service markets, pricing was often a key differentiator. This drove many service providers out of the marketplace. In addition, this price reduction tended to erode the value of the services, and ultimately weaken competitive positions, reduce switching costs, and decrease customer loyalty. In effect, stall a market's evolution.

To progress beyond this, service lifecycles need to transition from the Commodity quadrant into the Solutions quadrant on the Market Maturity Matrix, where service providers can begin targeting Market Movers as buyers by incorporating Methodology and Complete Service Solutions into their market plans.

Market Movers

The predominant buyers in Phase II Service Lifecycles are the Market Movers who are driven by operational efficiency, predictability, complete service solutions, and trusted sources. These buyers are accustomed to buying both direct from discrete service providers as well as from large service solution providers who integrate complementary services into single-source, or one-stop-shop service solutions.

Phase II buyers are more strategic than Phase I buyers. They want a sustainable operational advantage as a result of this service, not just a short-term, tactical advantage. This behavior is similar to product solution buyers in Quadrant III who want the assurance that the company and the service will be around for a while. These buyers also want the assurance that there are economies of scale in working with a single service provider and that they have a demonstrated ability to deliver this service predictably and consistently.

For marketers at this phase, transitioning to Phase II service markets offers interesting opportunities to package and brand methodologies, approaches, and processes for the delivery, use, management, or measurement of services. A marketer's ability to attract Market Movers is contingent on the ability to appeal to the buyer's operational needs, desire for complete solutions, strategic purchase value, and demands for guaranteed levels of service consistency and quality.

Methodology

To succeed in Phase II Service Lifecycles, service providers need to be able to scale their business, and offer complete service solutions that achieve the economies that drive profitability. To get there, however, service providers must pass an important milestone along the way by introducing a methodology or approach to delivering, managing, positioning, applying, or supporting their service that differentiates them within the market.

Service providers use methodologies in a variety of ways. In its purest form, a methodology is simply a proven, repeatable service method packaged into a discrete value proposition for target buyers. In service markets, methodologies are an opportunity to create differentiation and competitive advantage by emphasizing a company's unique competency into a formalized and scaleable process for delivering the service. In doing so, service providers can help buyers more clearly distinguish competitive offerings by "packaging" their approach to service delivery, quality control, management, support - just about anything that helps intersect their core competencies with key market needs.

First movers in Phase II service markets use methodologies in two important ways. First is for competitive reasons - to make processes, service levels, and results more predictable and reliable. This is a key decision criterion for Market Movers. Second is to create processes that are scaleable and repeatable for the service provider, which will enable them to grow as volume increases without sacrificing service quality.

Examples of this might include service packaging or "productizing" service delivery, implementation, training, reporting, call handling, legacy integration, service management, compatibility guarantees, service speed, or scaleability with a brand name, acronym, or graphic identity. Often, vendors at this stage will back up methodologies with Service Level Agreements to guarantee satisfaction.

Methodologies are timely in Phase II Service Lifecycles for a number of reasons. First, during Phase I of service lifecycles, service delivery is traditionally unique for each client, which had advantages at that phase. It allowed companies to treat each client implementation as a learning process to define and develop the core competencies and best practices needed in Phase II. Second, introducing a branded approach or methodology in Phase II appeals to the conservative nature of Market Movers with a credible and reliable process. Third, a structured approach or methodology also helps to streamline service delivery, which makes it faster and more scaleable. This, in turn, enables service providers to grow more quickly and better meet customer needs and timetables.

***Law of Perpetual Marketing**: Methodologies offer service companies a unique opportunity to concurrently create scaleable, repeatable processes, while creating competitive market distinction and advantage.*

When the market for outsourced web management services entered Phase II, although a flood of competitors entered this space, the first movers were the ones to introduce methodologies that enabled them to appeal to larger, more mainstream buyers with a proven, repeatable approach. Now, nearly every reputable web management firm uses a methodology for understanding their client, their client's market and the best approaches for site design, navigation, layout, etc. These methodologies follow a structured approach to discovery, analysis, recommendations, validation, graphic design and so on. This enables clients to better predict time, costs, and efforts necessary to complete their project. Thus, this structured approach breeds client confidence while creating a more repeatable means of delivering the service.

Service Solutions

Although methodologies and formalized approaches help service providers create competitive advantage and differentiation in this market phase, competition eventually will introduce its own methodology or approach that dilutes this advantage. Once this happens, methodologies become nothing more than a necessary cost to compete.

In the process, however, two critical steps occurred to prepare the market for continued growth. First, methodologies helped to attract Market Movers, which accelerated the growth and popularity of this market. Second, now that service providers have proven they can provide discrete services predictably well through their methodology, it becomes much easier to broaden service offerings and channels of delivery without compromising service quality. In essence, they can now offer complete Service Solutions using the same proven processes, approach and methodology used to deliver and manage their discrete services.

Complete service solutions include the convergence or integration of complementary services into a one-stop-shop for customers. Much like in product lifecycles, this can be done through mergers, acquisitions, strategic partnerships and alliances. But in service markets, where services are based on standardized technology, adding complementary services is also done through partnerships with service originators or technology companies on which the services were created. In turn, these complete service solutions provide conservative, operationally oriented, strategically-driven Market Movers with confidence and long-term stability. In service markets, service solutions can be offered through channel resellers such as service solution providers, or direct, if the service provider can scale to extend their range of services to compete effectively with service solution providers

In the example of outsourced web management services, a service solution would expand beyond site design and hosting to include services such as intranets, extranets, e-commerce, portal design for supply chains, e-mail hosting, firewall security, instant messaging, integrated Voice-Over-IP, Internet and broadband services, or Internet conferencing services. In doing so, service providers can expand their "share of customer" and quickly increase their strategic value to the client.

For marketers at this stage, creating complete service solutions fuel customer acquisition if these additional services can: 1) be integrated seamlessly into existing services and methodologies; 2) leverage company core competencies; and 3) support overall company positioning and goals. If these criteria can be met, service convergence is positive for customers, the market, and the service provider. Complete service solutions can also offer increased potential for customer retention and loyalty programs, including individualized cross-sell and up-sell opportunities for each client.

The danger with complete service solutions is the potential to drift away from company core competency, and sell only based on a complete solution. Marketers at this stage need to remember that service solutions *reflect* their core competency and *extend* their differentiation – they don't create it.

SERVICE LIFECYCLE PHASE III

One of our clients was an outsourced software application developer that had adopted a new methodology called Extreme Programming (XP). It allowed their developers to create software applications for businesses more quickly, more reliably, and with more flexibility than through traditional methods.

When we asked the executive team about their successes and failures, the President said, "It seems when we develop for telecommunications companies, we have the greatest success. We have more problems with manufacturing, financial services, and healthcare ."

We asked, "Is that because XP isn't appropriate for those vertical industries?"

"Not at all," the President said. "The methodology allows us to achieve the same results with the same benefits regardless of industry."

The VP of Sales chimed in. "The problem with those vertical segments is that our programmers don't always understand the business processes that are unique to those markets as well as they do for telecommunications. In fact, our sales people have the same challenge."

We said, "Then perhaps it's time to get domain expertise that will allow you to offer solutions unique for each vertical segment you are going after. In other words, perhaps you need to get closer to your customers by offering a "Healthcare XP Solution," a "Manufacturing XP Solution," and a "Financial Services XP Solution."

After going through the Market Maturity Matrix, and the keys indicators for Phase III Service Markets, the President said to us, "You're right. We need to look again at solutions that meet the specific needs of each of our target segments. But," as he turned his head to his executive team, "we're not going to be able to do this unless we align our sales and marketing, development, technical services, and support teams to back it up."

Service Lifecycle Phase III is the upper left quadrant of the Market Maturity Matrix, defined by the intersection of Niche Markets and New Technology, as illustrated in Diagram O.

Diagram O: Phase III Service Lifecycles

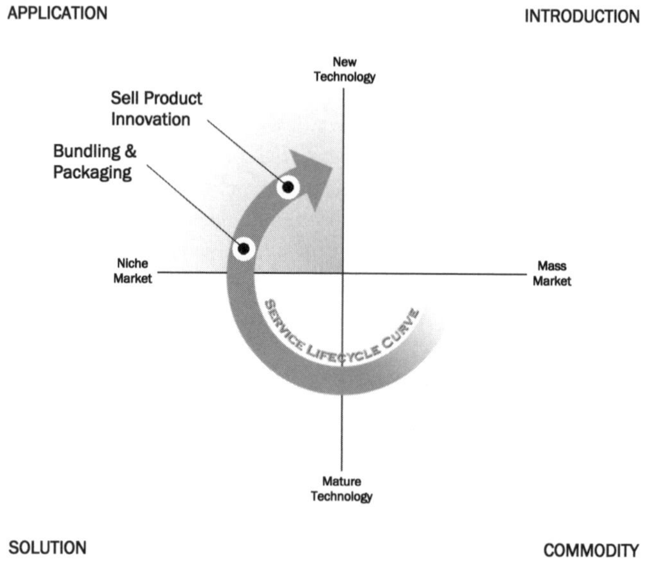

At the end of Phase II Service Lifecycles, service solutions emerged to meet the conservative, one-stop-shop needs of Market Movers. However, with little technology differentiation between their offerings, service providers are now becoming increasingly reliant on intangibles to create differentiation, such as customer satisfaction rates, brand recognition, and positioning. This makes business planning and forecasting more difficult, and challenges service providers to look for ways to create more tangible differentiation and higher margins. This drives the transition to Phase III.

Phase III service lifecycles are the Application quadrant of the Market Maturity Matrix. Much like the challenge product marketers faced when transitioning into this quadrant, service marketers face similar challenges in tying to target market segments with unique applications or solution components. For service providers, one of the most successful ways of entering Phase III Service Lifecycles is by creating service Bundles and Packages for niche markets based on specific industry or business needs. Creating service bundles and packages create barriers to entry for competitors by making the service a more strategic component of key business initiatives. In turn, bundling and packaging designed to solve specific business problems can also increase switching costs and improve customer loyalty.

Although bundling and packaging add strategic value to clients, they can go one step further. By leveraging new and emerging product technologies, service providers reach the final stage of service markets, where services can be used to Sell Product Innovation.

***Law of Perpetual Marketing**: Services can actually sell new product innovations.*

One of the reasons Phase III is appealing to service providers is pricing and margin opportunity. Unlike product lifecycles where prices of core products and technologies continue to drop over the life of the market, technology service pricing and margins have the opportunity to increase as service markets transition into Phase III. This is because, unlike the Market Movers in Phase II looking for operational improvement, Phase III buyers - Technology Movers - are looking for competitive advantage and unique applications for the service in their business. They want to buy services packaged in such a way that provides them with a catalyst for business improvement.

With these higher profit margin opportunities, Phase III is an ideal time to build channel partnerships to reach specific market segments with unique applications or solution components based on the service. By leveraging the resellers' specific domain expertise, service providers can use bundling and packaging as well as new product innovation to target markets beyond their core competency. Channel partnerships at this phase can include VARs, service solution providers, product companies - just about anyone who shares the desire to extend client value and has access to a customer base.

Technology Movers

To reach Phase III Service Lifecycles, and the rewards of success in this phase, service providers must transform their business model, pricing model, service packaging, and delivery model to accommodate the changing needs of buyers at this phase.

Buyers in this phase are predominantly Technology Movers. These buyers are most often looking at services as components to complement existing initiatives that will act as change agents for their business, and are primarily driven by a desire to create competitive advantage in their marketplace. These buyers place value on domain or application expertise in their industry or business function, and hence like to work with service providers who understand how services can solve particular problems.

Similar to product buyers in this quadrant, service buyers will expect domain expertise and application knowledge within their industry or business function that helps them use this service for competitive advantage. They will also demand specific value-add beyond what can be achieved through repackaging alone. This might include integration with certain proprietary legacy systems, unique user interfaces, new capabilities or functionality, certain types or reports or metrics, special customer or industry-related web portals, or certain levels of quality, performance, or personalized service unavailable from complete solution vendors.

Bundling & Packaging

Transitioning into Phase III requires an ability to deliver services to more target market segments with a more narrowly focused value proposition. In essence, it is target marketing. With services, the best first way to deliver this is through bundling and packaging of service components along with key elements that make it unique for each customer or niche market segment - such as complementary services, new technology, or custom services. Where in Phase II service lifecycles, the goal was to expand the range of services and offer complete and flexible service solutions, Phase III buyers want service solutions designed to their unique needs.

As an example, lets take a look at the cellular phone service market. In this market, transitioning to Quadrant II included packaging monthly services for discrete type of users, where the service providers could increase monthly minimum plans by offering value specific to their target audiences. This included such packages as family plans to share multiple phones, combined wireless voice and text messaging plans, free weekend calls, or wireless web plans.

Another example is the outsourced call center market. This market reached the service solution stage in Phase II with vendors such as Teletech who offered consistent, predictable service that could readily be expanded into complete service solutions. For this market to progress into Phase III service markets, it took smaller and more nimble call center service companies to target specific market segments with customized services that leveraged emerging technologies such as Voice-over-IP, integrated web and telephone customer interaction, and wireless web.

Sell Product Innovation

Once services have matured to this point, the final stage for services is to use these services to sell new product innovation. New product innovation can include hardware, software, content, access to information, or even new service products – just about anything that can strategically extend the value of the service or service delivery. One of the keys to new product innovation is in enabling service providers and their channel partners to leverage new and emerging technologies that extend the use and value of the base service.

To continue the example of the cellular phone service market, when you sign up for a cellular service plan, a basic phone is usually included for free. These are good phones that just a few years ago would have cost you well over $100 in addition to the service plan. But today, they are free in exchange for extended service contracts and, in most cases, are sufficient for an average cellular user. However, to use some of the more advanced cellular services such as wireless Internet, text messaging, and data services, you need to buy the $300 phone, and sign up for a new, more expensive service plan. This enables service providers to drive demand for higher priced, higher margin phones and services that leverage emerging technologies. In the process, it strengthens the relationship between cellular services and product companies by creating continuous demand for both products and services.

Another example is the Palm VII from Palm Computing. Before the Palm VII, data exchange between PDAs and other devices such as PCs was only available through docking stations or special cables. Service providers used the Palm VII as an opportunity to extend their service plans by creating wireless plans specifically for the Palm VII, led by applications such as wireless e-mail, stock tickers, news updates, and more. These applications extended the value of the service by allowing existing users to extend their account to accommodate these new devices and their limited data storage and screen display capabilities.

Section II: Applying Perpetual Marketing

6

Ready, Fire, Aim: Creating Market Focus

One of the most important benefits of *Perpetual Marketing* using the Market Maturity Matrix is that it provides focus. Focus is important to every company, of every size, in every industry, at every market phase.

In technology markets, even the largest companies cannot be all things to all people. Through focus, companies are able to rank and target market segments, differentiate from competition, create brand identity, and prioritize internal resources and initiatives to leverage and reinforce a company's core competency and brand.

> ***Law of Perpetual Marketing**: The definition of focus is "not being all things to all people."*

When helping our clients create focus or priority, a red flag we look for is when their current mission or positioning statement reads something like this:

> "Our mission is to automate business processes inside and outside the firewall by providing an asynchronous messaging layer with automatic queuing built on open standards which provides guaranteed reliable and completely flexible, dynamic, scaleable distributed processing with built-in security all centrally managed ensuring complete connectivity between front office, back office, eCommerce, custom, legacy and mainframe systems. In real time over the Internet."

Positioning statements like this confuse employees, customers, investors, media and analysts. A better version of this positioning statement might read more like this:

> "For companies needing improved Internet-based messaging, we guarantee the most reliable and secure messaging solution to provide connectivity between new and existing corporate systems."

This second version is not only more concise and understandable for external audiences, but it provides focus and priority for internal audiences.

Focus Doesn't Mean Exclusionary

Some executives balk at market focus as exclusionary. They are concerned about turning away opportunistic business. They want to remain flexible, nimble, and appeal to a large audience. Opportunistic business is important at each market phase, but must be treated as such. In other words, opportunistic business should not distract or redefine outbound efforts or internal priorities. It should be recognized and pursued for what it is – a one-off revenue opportunity – not a new business focus.

Focus is simply an approach to prioritizing outbound efforts and internal resources. Just because you focus on a certain market segment does not mean you have to turn away business from another segment. Instead, focus gives you the flexibility to be selective in determining which opportunities outside your focus to pursue.

For instance, in the example above, this company's target market is companies that need the most reliable and secure Internet-based messaging solution. However, through the normal course of business, they may attract isolated prospects more interested in their firewall capabilities or scaleability. In this case, they may not turn away these opportunities, but it also does not mean that they should change the company positioning statement to emphasize firewall or scaleability.

The Market Maturity Matrix helps focus marketing efforts to attract buyers best aligned with your current ability to deliver, and in turn, improves the quality of leads generated, eliminates extraneous activities, preserves precious marketing dollars, and minimizes wasted efforts.

7

Look Before You Leap: Market Entry Strategies

Far too often, technology companies enter markets without defined, achievable market goals or a unified strategy. Instead, they enter markets with only financial goals and a strategy to sell as much as possible to whoever will buy it. As a result, these companies look more like a three-ring circus with sales, development, and marketing each performing flawlessly, but with little activity in common.

Market entry strategies define common goals and objectives that leverage a company's core competency, while maximizing timely market opportunities. They set market and operational goals that give each department metrics by which they can track and measure their contribution to the united effort.

Successful market entry can be made at any time in technology markets. However, with each progressive market phase the cost to enter and compete goes up. The cost to deliver products and solutions; establish channels, brands, partners, and alliances; gain analyst recognition; and acquire customers increases significantly as markets mature. Hence, the more mature the market, the higher the cost to enter and create parity in these areas.

Making Strategy "Sticky"

A strategy is only good if it is internalized and made part of the daily lives of employees. For this reason, strategies need to be made "sticky" so employees can easily identify and remember key aspects. One way to do this is to use code names or phrases for the strategy, which can be either fun or serious. These code names or phrases help to rally workers around a common goal or mission and keep everyone focused on a common objective.

Jack Welch, the famous leader of General Electric, used to say "Keep it Simple" as a mantra to his people. This simple statement was intended to have broad implications across the organization. It implied that everything General Electric did from product design, to contract writing, to marketing materials should be kept simple and straightforward. Simply stated, it worked.

MARKET ENTRY STRATEGIES BY MARKET PHASE

Selecting the best entry strategy depends on a number of factors including:

- The timing of your entry – at what phase in the market's lifecycle
- Order you enter the market – first, latest
- Your company size and resources – versus direct and indirect competitors

Diagram P identifies some commonly used market entry strategies, organized by market phase. Although many of these strategies can be used across a variety of phases and a market entry can often include multiple strategies, this diagram makes it easy to align entry strategies with key market opportunities in each phase.

Diagram P: Common Market Entry Strategies

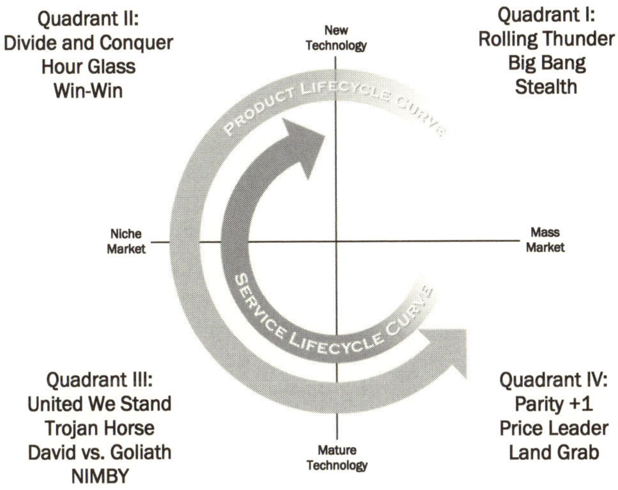

Quadrant I

Being first to market is important, but not essential for achieving ultimate market leadership. It's true that companies first to market benefit from expanded press and media attention. But companies first to market also must bear the cost of educating the market. They must act as the evangelist for the category of technology before they can begin to distinguish their particular implementation of the technology. For this reason, it isn't always the first company to market that ultimately wins a market category. Three of the most common ways to enter markets in Quadrant I are Rolling Thunder, Stealth, and Big Bang.

> *Rolling Thunder* is a conservative entry approach that builds on incremental successes to progressively expand. It allows companies to test the market for adoption indicators, competitive strengths and technology acceptance. The advantages of this approach include smaller start-up capital, growth based on success of previous efforts, and doesn't outpace a company's resources or ability to deliver or service customers. The disadvantage of this approach is that in fast moving markets or for companies without a significant technology lead, this approach takes too long to build up market momentum and market share.

Stealth introductions are the ones you don't hear much about - by design. Companies use this approach to selectively target outlets to disseminate information and news. This method is good for companies looking for funding or strategic partners, those still building out infrastructure, or those in final stages of product testing. The advantages of this approach are that it keeps you below the radar screen of most of your competitors, which protects your competitive advantages, but still makes you accessible to potential investors and partners. The disadvantages of this approach include a tendency for competitors to discredit a stealth company with claims of "vapor-ware" or an inability to launch, and the potential to miss key market opportunities.

Big Bang is for companies with significant financial resources who want to rapidly move their product through the product lifecycle. The advantage of this strategy is that it can propel a company immediately into the market leader spot in your industry, and establish the company as the de facto standard in the market. The disadvantage is that if it's an industry not well defined, has few barriers to entry, or lacks latent customer demand, much of your capital expense to build awareness may not translate into revenue, and will help competitors leverage your investments in market education.

Quadrant II

In quadrant II, market entrants can capitalize on the education and development of their predecessors while finding new and unserved market segments and applications for the technology. For product companies, this includes concentrating on building and leveraging domain expertise, channel access, or product applications to create differentiation and positioning. For service providers, this includes the ability to transition their complete service offering into applications and products tailored to specific customer or market segment needs. Market incumbents need to be sure to define a unique and defensible market position and resist urges to react to new entrants by chasing market segments, features, or positions that don't reflect the company's core competency. Three of the most common ways to enter markets in Quadrant II are Divide and Conquer, Hour Glass, and Win-Win.

Divide and Conquer is an approach based on niche market segmentation and selection. This approach is common when competitive products lack significant feature/function differentiation, but offer significant opportunities for creating value by leveraging domain or application expertise related to the product to solve unique problen market segment. The advantage of this approach is unique and defensible positioning opportunities for companies that can build, buy or partner domain expertise in target market segments. The disadvantage of this approach is that it can limit long-term growth if the model for success in one niche market is not transferable to subsequent markets.

Hour Glass is an approach for companies that struggle to select a single niche market as a point of market entry. This approach allows companies to 'test the waters' in several markets concurrently, then narrow their focus as they gain experience and customers. The advantage of this approach is that companies don't limit their market entry to a single niche market. The disadvantage is that companies have to make a larger initial capital investment in marketing, sales, and domain expertise equally across a range of markets while they build revenues.

Win-Win is an approach for companies that enter the market specifically targeting partners and resellers as their sales and delivery channel. This approach helps define messaging and positioning that aligns with target partner and reseller goals and target customers. The advantage of this approach is that it minimizes time and capital investments in broad marketing and sales efforts by focusing efforts on a select number of resellers who recognize the win-win opportunity of representing your product. The disadvantage of this strategy is your dependency on these channel partners for marketing, selling and potentially servicing your customers, raising issues such as customer ownership and channel loyalty.

Quadrant III

In quadrant III, companies can look to integrate complementary companies, applications, and services faster than their competitors. Whether done so through mergers, acquisitions, or convergence, this creates opportunities to establish new buyer expectations by redefining decision criteria and setting higher demands for what complete solutions include and the results they deliver. New product entrants have an opportunity to introduce product solutions, new distribution channels, new pricing models and standards, and new delivery techniques. New service entrants have an opportunity to establish new buying decision criteria that offers a predictable and reliable methodology or approach to service delivery and management.

Incumbents in this phase have an advantage in that they should understand the market and the technology better than new entrants. Incumbents should be better connected, and quick to respond to new entrants by leveraging existing partnerships, alliances and industry connections to converge the "best of breed" components into the best solution in the market. Some of the most common ways to enter markets in Quadrant III are United We Stand, Trojan Horse, David Versus Goliath, and NIMBY (Not In My Back Yard).

United We Stand is a partnership or alliance approach. In this approach, multiple companies with complementary products and solutions join forces to compete against larger, more established solution providers. The advantage of this approach is that it unites the resources of multiple complementary vendors into an integrated value proposition for the market. The disadvantage is the perception of product integration, multiple points of accountability, and multiple points of contact.

Trojan Horse is an approach for creating a stronghold inside the customer using a single point of entry. This might include a loss leading product or service, or a time sensitive offer. Then, once a relationship is established, you leverage this to extend share of customer or displace the incumbent. The advantage of this strategy is that it exploits a company's strengths and creates specific opportunities to create customer satisfaction. The disadvantage is that it takes longer to build up economies of scale; it often requires the same sales and marketing effort to make smaller initial sales.

David versus Goliath is an approach for taking on large incumbents by specializing in one particular aspect of the solutions' functionality or use. This approach is often used by companies who have distinct differentiation in how a solution is sold, implemented, managed, or supported, or have differentiation through unique capabilities or functionality. The advantage of this approach is that it maximizes differentiation in a market that may be nearing commodity. The disadvantage is that the differentiation must be compelling enough to establish value, and it must be sustainable long enough to so it can't be easily copied by the competition.

NIMBY (Not In My Back Yard) is a defensive position. It is a strategy used by incumbents who need to enter complementary markets to protect their customer base from smaller or more nimble competitors. The advantage of this approach is that it forces companies to emphasize customer service and satisfaction to minimize cannibalization of their customer base. The disadvantage is that since it is a defensive position, it can seem reactive to investors and shareholders that prefer to see more growth-oriented programs.

Quadrant IV

In quadrant IV, the establishment of standards enables cost leaders and operationally efficient companies to thrive. For product companies looking to enter markets at this stage, the two most important criteria for success are 1) a more efficient sales and delivery operation and 2) access to operationally and cost efficient suppliers. For service companies, the cost to enter and compete in this phase is generally low, but at a minimum should include 1) domain or application expertise, 2) "parity plus one" functionality or service differentiation, and 3) access to the technology source.

When Canon wanted to enter the market for digital cameras, its only barriers to enter this market were cost-efficient manufacturing and a sales process. Since it already had established channels and corporate brand value, its credibility for expanding into digital cameras beyond film-based cameras was nearly automatic with the channel and with consumers.

For incumbents in these markets, being early to recognize the operational changes necessary to sustain a profitable business model in this market phase is key to succeeding. Also, be sure to secure distribution and delivery channels, and strive to create barriers to entry from new competitors. Two of the most common ways to enter markets in Quadrant IV are Parity+1, Underdog and Land Grab.

Parity+1 is an approach for companies that can more rapidly and consistently introduce continuous innovations to extend the life of the product versus the competition. The advantage of this approach is that continuous product extensions provide clear direction and focus for product and marketing teams. The disadvantage is that the "parity plus one" innovations usually have short windows of opportunity that must be monetized very rapidly.

Price Leader is an approach to undercut existing competitors and become the cost leader in the market. This approach is favorable for companies that have two key advantages: an operational or cost advantage in manufacturing and production, a unique approach to sales and delivery. The advantage of this strategy is the ability to capitalize on established market demand. The disadvantage is that a cost advantage is usually only temporary, and can usually only be used as an entry strategy. This strategy requires a follow-on strategy to transition away from price into another form of differentiation or brand development.

Land Grab is primarily for service companies with significant financial resources who want to rapidly penetrate a geographically dispersed market. The advantage of this strategy is that it builds momentum and recognition very rapidly. The disadvantage is that if it's an industry not well defined or has few barriers to entry, much of the capital expense used to build awareness may not materialize and will ultimately help competitors leverage your investments in market education.

8

May The Force Be With You: Market Transitions

When market lifecycles transition from one phase into the next on the Market Maturity Matrix, it is both disruptive and opportunistic. When markets transition, buyer behaviors change as do their drivers, needs, and expectations. This often requires changes to business models, pricing methods, distribution methods, and development or engineering methods. Most of all, it requires changes to marketing practices.

Market transitions are ideal times for market entry. This is true at nearly every market transition point. Market transitions offer unique opportunities to redefine buying decision criteria by leveraging the emerging needs of buyers in the new market stage.

***Law of Perpetual Marketing**: Market stage transitions offer the best opportunities to enter markets or redefine buyer decision criteria.*

An example of this is the network appliance market. In the late 1990s, as the explosion of the Internet and network applications such as e-mail and web site hosting grew within the small business market, there was a growing demand for dedicated servers to perform these new functions. With the premise that existing network servers, application servers, and mini-computers were a mismatch for the needs of these new applications within the small business market segment, a new breed of specialized servers emerged to simplify handling of these applications and content. They were dubbed network appliances. Several competitors emerged in this market, including one company called Cobalt Networks.

As the network appliance market transitioned from the Introduction phase to the Application phase on the product lifecycle curve (Quadrant I to Quadrant II), Cobalt Networks used this market transition to create a unique position and value for its specialized server appliance that changed the buying criteria for this market. Cobalt Networks didn't promote MHz or number of processors; rather their unique combination of consumer class functionality, reliability, and pricing in an appliance that offered a completely new way of handling e-mail, web content, and other specialized networking functions for the small business market. This was appealing to the buyers in the Application phase who valued the competitive advantage of applying these products to change existing methods, and solve specific business problems. In just three years after startup, Cobalt Networks grew to nearly $80 million in revenues, and was summarily acquired by Sun Microsystems in late 2000 for $2 billion.

Tips for Market Incumbents

Market transitions are perfect times for incumbents to be ambushed by a new competitor. Market entrants without legacy customers and products can often act swiftly and work hard to redefine market decision criteria, expectations, prices, or product/service capabilities. This can often times force incumbents into a reactive mode where they quickly work to establish parity with these new entrants - often at the expense of existing product roadmaps or development plans.

Using the Market Maturity Matrix, market incumbents can better predict and prepare for these transitions. This minimizes reactionary management and improves the quality of decisions used to respond to changing market dynamics. Using the Market Maturity Matrix, marketers can better prepare their company with contingency and eventuality plans that account for market and competitive events.

The story of Novell is a good example of a company victimized by market transition. Once the dominant name in Local Area Networks

(LANs), Novell NetWare was a very good product, had an expansive reseller network, and one of the best brand names in the technology industry. It also had a complete aftermarket dedicated to training and certification, products enhancements, users groups, trade shows, publications, and more. For years, Novell was the first mover in its market and led the industry with key innovations in technology, distribution, and installed base. However, as the Local Area Networking market transitioned from the Solution phase on the product lifecycle curve into the Commodity phase (Quadrant III to Quadrant IV), it appeared that Novell was unprepared to embrace the commoditization of this technology. As the market matured into the Commodity phase, Microsoft used this market transition to change the buying criteria for the LAN market by embedding TCP/IP and file and printer sharing components directly into the Windows operating system. This created a de-facto standard in LAN connectivity that forced Novell to follow a different path.

Tips for New Entrants

Market transitions are ideal for disrupting the status quo. By capitalizing on the changing needs and motivations of the market, new entrants can appeal to the changing purchasing and adoption habits, messaging strategies, communication vehicles, and delivery channels that best align with the emerging market's characteristics. This enables new entrants to capitalize on the education done by their predecessors and build on this foundation with differentiation based on enhanced capabilities, functionality, performance, quality, reliability, or anything that intersects latent market demand with your company core competency.

In these situations, the Market Maturity Matrix can be used to help track and anticipate market transitions and better understand an emerging market's needs. This leads to better decision-making, better entry strategies, and more focused launch activities.

9

Market E-motions: Changing Buyer Characteristics

Buyer characteristics change across market lifecycles based on different emotional and conditional drivers. The Market Maturity Matrix helps by creating clear and identifiable distinctions between buyer behaviors at each market phase.

> ***Law of Perpetual Marketing**: Buyers at different market stages have very different buying behaviors.*

Diagram Q distinguishes buyers by quadrant on the Market Maturity Matrix. A complete list of buyer characteristics has also been summarized at the end of this chapter in Table D.

Diagram Q: Buyers by Quadrant

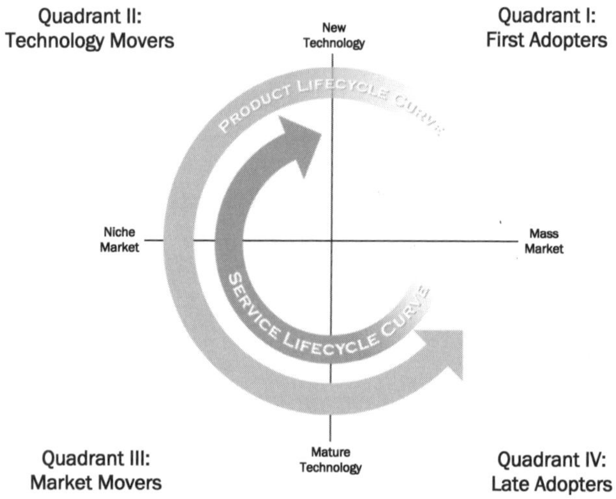

Quadrant I: First Adopters

Quadrant I is the upper right quadrant of the Market Maturity Matrix. This quadrant is defined by the intersection of new technology and mass markets.

Buyers in this quadrant are called First Adopters, and they are truly unique in their buying behavior. This can make this stage deceiving for product manufacturers, as these are usually the first enthusiasts who take an interest in your technology. First Adopters are visionaries who look to technology as a catalyst. They gravitate toward discontinuous innovations like sailors to the Siren's song. By nature, they are information hungry and are always in search of the "next big thing". For business-to-business products, these buyers usually buy more from an individual need than a business need, but still recognize that these technology components can bring them a market advantage. They tend to be emotional in their purchases, which makes them more tactical in nature, more financially aggressive, and more risk prone than buyers at any other stage. First Adopters tend to interface with non-vertical peers, often meeting and discussing technologies like kindred spirits drawn together through a common passion for technology. They almost always purchase direct from manufacturers because they want to have an influence over the development and evolution of the technology.

For instance, when wireless devices (i.e., cellular ph[illegible] were first introduced, buyers of these products had very [illegible] in terms of motivation, application and use, risk tendencies, exp[illegible] and price sensitivity than buyers of wireless devices now. Early cell phone adopters were primarily business and service professionals such as executives, sales representatives, doctors, lawyers, and others whose work required that they be constantly accessible. Now, with price, service, and packaging of wireless devices fairly standardized and extremely affordable, just about everyone owns a cellular device.

Quadrant II: Technology Movers

Quadrant II is the upper left quadrant of the Market Maturity Matrix. This quadrant is defined by the intersection of new technology and niche markets.

Buyers at this stage are called Technology Movers. They are motivated by conditional or competitive reasons, and typically are looking for solution components to use as change agents. For this reason, buyers at this stage are more strategic in their purchases, moderately aggressive financially, and are willing to accept a certain degree of risk to implement these products. These buyers are influenced by technology trends and look to market influencers and industry organizations for decision validation, such as analysts and industry opinion leaders. Buyers in this quadrant tend to purchase either direct from the manufacturer or through specialty providers who offer domain expertise in the application of this technology, the business problem being solved, or the functional area of the business this application addresses.

In Quadrant II, product lifecycles take their first steps from technology to "productization." As these products take shape and are packaged for buyers in discrete market segments, they take the shape of applications designed to address specific inefficiencies in current approaches or methods. Buyers in the phase look to these products to bring them a distinct competitive advantage.

For service markets, Quadrant II is the final phase of service lifecycles. Buyers at this stage are less interested in the economies of scale that attracted their predecessors. Rather, they are motivated to leverage this service in unique ways to create competitive advantage. For this reason, "one-stop-shop" messaging can often miss the mark. Instead, buyers at this stage are looking to service providers to bundle service "packages" or "applications" that combine select services with new technology products or complementary services to meet their unique needs.

Quadrant III: Market Movers

Quadrant III is the lower left quadrant of the Market Maturity Matrix. This quadrant is defined by the intersection of mature technology and niche markets.

Buyers in this quadrant are called Market Movers. They are looking for total solutions from both product and service vendors. These buyers are conservative in nature and want selection and comparison to ensure they select the solution that best suits their needs, at the best value. These buyers are typically looking to use solutions to automate existing processes. They are financially conservative and are influenced by market trends and trade organizations. These buyers tend to buy through large implementers or direct from manufacturers who can deliver complete solutions. Purchases at this stage are more strategic than tactical, and tend to be operationally driven.

Product buyers in this quadrant are looking for technology and technology-based products to bring them competitive parity. These buyers are trying to keep pace with market, but without investing in the learning curve and integration costs of their predecessors. For this reason, they tend to prefer integrated product solutions, including the professional services to rapidly implement and benefit from this solution.

Service buyers at this stage are more sophisticated and demanding of service provider, are looking for more predictability and assurance from the service as well as the provider. They tend to favor service providers with an established brand or channel partner recognized for consistent and reliable service delivery and management.

Quadrant IV: Late Adopters

Quadrant IV is the lower right quadrant of the Market Maturity Matrix. This quadrant is defined by the intersection of mature technology and mass markets.

Buyers in this quadrant are called Late Adopters. Although they may be early adopters in service markets, they are called late adopters because they buy technology in a mature state. Markets in the commodity stage are identifiable by buyers looking for market parity; nothing too risky, nothing too new. For this reason they tend to delay purchase decisions until they become a necessity. They are financially very conservative and tend to use technology purchases more for tactical gain than strategic gain.

Product buyers in this quadrant look for mature, turnkey, and embedded solutions. They place value on the brand of the product they are purchasing, the brand of the channel they are buying through, and the value-add of the channel they buy through. For this reason, they tend to buy through mass channels or contracted suppliers. Their purchases at this stage tend to be financially driven.

Service buyers in this quadrant are typically looking at services to reduce product implementation risks or the risk of continuing with current methods or solutions. These buyers look for services based on proven, mature technology, and tend to buy direct from service providers or large service implementers who offer economies of scale, expertise in the service technology, and fast time-to-market.

r Behavior by Quadrant

Buyer Behavior	
Quadrant II: Technology Movers • Looking for competitive advantage (lead industry) • Moderately aggressive financially • Risk acceptance • Use products and services as change agent for business • Purchase direct or from specialty resellers with domain expertise in market segment or business function • Looking for product and service applications unique to their business or market • Influenced by industry organizations and technology analysts • Want personalized service and attention • Conditional or competitive purchase	**Quadrant I: First Adopters** • Looking for market advantage • Looking for new technology and discontinuous innovations • Financially aggressive • Risk prone • Use new technology to gain business or individual recognition • Buy direct from manufacturer (want to have influence on product direction) • Influenced by peers and industry visionaries • Want information and education • Want to share in the vision • Emotional purchase (tactical)
Quadrant III: Market Movers • Looking for competitive parity (keep pace with industry) • Looking for complete solutions • Financially conservative • Risk averse • Use products and services to automate or improve existing operations or processes • Purchase through large implementers or direct (if vendor can deliver complete product or service solution) • Influenced by market trends, trade organizations, and technology analysts • Want competitive selection and comparison • Operational purchase (strategic)	**Quadrant IV: Late Adopters** • Looking for market parity • Look for mature, turnkey, or embedded technology or solutions • Financially very conservative • Risk avoident • Product buyers use new technology to replace outdated or existing technology • Looking for expertise in the service technology, and fast time-to-market • Purchase through mass channels or contracted suppliers • Purchase direct or through large service implementers • Influenced by channel, brand recognition, perceived value, and availability • Want availability and reason to choose your brand

10

The Laws Of Perpetual Marketing

The study of Perpetual Motion is famous for its "Laws of Motion," such as Sir Isaac Newton's Third Law of Motion: "for each action, there is an equal and opposite reaction". These laws are the foundation of physical science, and have led to discoveries that ultimately helped break the speed of sound and put men on the moon.

Similarly, *Perpetual Marketing* has its own "Laws of Perpetual Marketing" as summarized in Table E. These laws may never be as famous as Sir Isaac Newton's laws, but they offer snapshots of *Perpetual Marketing* wisdom to help you make better decisions with more predictable results.

'erpetual Marketing

Laws of Perpetual Marketing

1. Perpetual Marketing is simply "the right marketing at the right time."
2. The Market Maturity Matrix acts as both a compass and a timepiece for technology markets.
3. Technology market dynamics are distinct and predictable across market stages.
4. The two most influential forces driving technology markets are technology evolution and market adoption.
5. Technology innovations can be continuous or discontinuous. Know the difference.
6. Mass markets buy tactically and niche markets buy strategically.
7. "Skipping" an evolutionary market stage is rarely desired or rewarded.
8. Technology is only an enabler. It's not a product until it serves a market.
9. Competition is necessary and should be embraced.
10. Segmenting and targeting markets is difficult but necessary.
11. Product solutions are the convergence of complementary applications and professional services into a single, integrated purchase decision for buyers.
12. For commodity markets, brand is king.
13. Service markets invite many competitors due to few barriers to entry and attractive recurring revenue streams.
14. Methodologies offer service companies a unique opportunity to concurrently create scaleable, repeatable processes, while creating competitive market distinction and advantage.
15. Services can actually sell new product innovations.
16. The definition of focus is "*not* being all things to all people."
17. Market stage transitions offer the best opportunities to enter markets or redefine buyer decision criteria.
18. Buyers at different market stages have very different buying behaviors.

11

Put Your Money Where Your Market Is: Translating Strategy Into Tactics

As mentioned earlier, the Market Maturity Matrix is more than just a theoretical model; it is an applied model. This means it not only provides strategic insight, but also provides tactical considerations for executing on your strategy.

To this end, many of the tactics included in this book have been distilled into the following tables. These tables highlight specific marketing considerations for each stage of product and service lifecycles, and provide a quick reference to help focus and prioritize budgets, resources, and activities to create marketing plans that result in "the right marketing at the right time."

These tables are formatted into quadrants to align with the market phases for product and service lifecycles.

Marketing Tactics Versus Infrastructure

The following tables focus on marketing tactics, not on marketing infrastructure. Marketing infrastructure includes fundamental marketing materials such as company identity (logo, stationery, etc.), product descriptions (data sheets), and press kit materials (company, technology, and management descriptions, etc.). Marketing infrastructure is usually considered the minimum marketing materials necessary to be in business.

The tables below focus on marketing tactics or activities used to create and qualify leads, establish competitive positioning, build brand, develop channels, support sales organizations, and drive product plans. Tactics account for the peaks and valleys in budgets. Tactics comprise the programs and initiatives created to meet business and market goals.

MARKETING TACTICS BY MARKET PH PRODUCT LIFECYCLES

Market Development

As product lifecycles mature, each market phase offers unique opportunities to maximize market penetration and growth. This is significant for making key decisions related to market focus, selection, and expansion.

Market Selection and Expansion	
Phase II • Marketing activities designed to be targeted and participative • Focus on capturing data on repeatable implementations and consistent uses of the product in targeted market segments • Understand how the product is being applied to solve business problems and how users are quantifying savings or improvements • Primarily direct approach, but may also include indirect marketing through select channel partners and resellers	**Phase I** • Marketing activities designed to be interactive and educational • Programs should be designed to capture and learn from your experiences • Involve yourself in customer implementations and use • Capture and analyze experiential data in preparation for market transition to Phase II • Keys to this phase is availability • Leverage indirect channels and mediums • Leverage public relations and network and viral marketing to reach potential first adopters
Phase III • Integrate direct and indirect programs in equal proportions to maximize exposure, reach, and number of impressions • Focus on building brand and consistency of message and position • Outbound direct marketing prioritized by defined customer sets • Indirect marketing activities identified to complement direct marketing activities by overlapping target market sets and increase the number of touch points to prospects	**Phase IV** • Maximize exposure through customer preferred Points of Purchase (POP) • For direct sales organizations, this will include many of the same integrated techniques used in Phase III • For indirect sales organizations, participation in existing indirect channel and partner programs such as POP direct mailers, telemarketing, advertising, and on-line marketing and merchandising

Market Focus	
Phase II • Targeted (based on domain expertise or specific/unique knowledge) • Ability to segment and focus on select target markets • Selectively expand markets by adding domain expertise and replicating value proposition • Form alliances and partners with specialty and/or domain expertise or competence in complementary market segments • Product convergence through mergers and acquisitions (competitive and complementary)	**Phase I** • Broad market availability • Experimental • Recognize and quantify new market opportunities based on technology lead or specific competence • Attract early adopters or visionaries
Phase III • Mainstream market buyers • Direct and expanded channel programs expansion • Ability to serve customers of varied back-grounds and needs with consistent and predictable solution • Channel expansion • Co-marketing (alliances and partners) • Differentiation based on value-added service and customer experience • Customer referral programs • Increase emphasis on brand development	**Phase IV** • Mass market (indirect – leverage channels for market saturation; direct – large, corporate accounts and new, emerging and unserved markets) • Branding and positioning • Channel partnerships • Co-marketing • Extend product life as embedded technology

Lead Development

Without priority and focus, lead generation programs can ofte[illegible] laundry list of activities. By using the Market Maturity Matrix, [illegible]keters can prioritize and focus direct and indirect marketing activities such as trade shows, direct mail, telemarketing, on-line marketing, and advertising to align with the activities that typically yield the greatest return at that phase of the market lifecycle.

Trade Shows, Seminars, and Conferences	
Phase II • Shows and conferences targeted at core market segments • Shows targeted at channel partners and resellers • Technology industry shows and conferences (either alone or as part of partner booth) • Partner shows and events • Application-driven educational seminars (either hosted alone or with partners)	**Phase I** • Technology focused shows • Industry show featuring new technology announcements and introductions
Phase III • Large industry shows and conferences targeted at channel partners and end-users • Trade organization shows and events • Partner shows and events • Topical seminars (either hosted alone or with partners)	**Phase IV** • Large industry shows and conferences • Exhibit with resellers

Direct Marketing	
Phase II • Direct mail and telemarketing to prospects within target segments (either alone or with partners) • Application-focused (not technology-focused) messaging • Calls-to-action used to qualify prospects out • Partner programs to supplement target market focus	**Phase I** • Selective direct marketing to known technology enthusiasts (technology-focused messaging) • Selective direct marketing with industry influencers
Phase III • Direct mail and telemarketing to prospects across broad range of markets (may still have vertical applicability) • Solution-focused messaging (key on competitive differentiation) • Calls-to-action used to qualify prospects in Partner programs to extend geographic reach and delivery options	**Phase IV** • Brand-focused messaging • "Parity plus one" messaging

Advertising	
Phase II • Select vertical or target market publications (alone or with partners) • Select technology industry trade publications • Select trade organization publications • Target strategic technology buyers • Select channel publications • Application-driven messaging • Use to build credibility via proof and endorsement	**Phase I** • Select sections of technology industry trade publications (i.e., 'New Products' section, 'What's Hot' section)
Phase III • Technology industry trade publications • Select business/consumer publications • Industry trade organization publications • Target strategic solution buyers • Solution-focused messaging • Key on competitive differentiation • Use to build brand and reinforce positioning	**Phase IV** • Point-Of-Purchase (POP) for retail • Channel circulars and Run-Of-Press (ROP) • Technology industry trade publications • Business/consumer publications • Radio (based on demographic) • Outdoor (based on geography) • Brand-focused • Parity plus one messaging

On-line Marketing	
Phase II • Second-generation web site • Targeted e-mail programs (ideally integrated with off-line programs) • Select banner ad placements and offers based on target market segment • Trade organization chat rooms and news group postings • Initial e-commerce for customers and partners • Web seminars based on applications by target segment	**Phase I** • First generation web site • Education-oriented • Chat rooms and news groups focused on technology discussions • Search engine listings
Phase III • Advanced, interactive web experience for prospects, customers and partners • Banner ad placement and offers to expanded markets • E-mail based newsletter for customers • Trade organization chat rooms and news group postings • Web seminars based on industry solutions and customer experiences • Partner web site listings • Advanced e-commerce for customers and partners • Self-service support	**Phase IV** • Advanced, interactive web experience for prospects, customers and partners • Banner ad placement and offers to expanded markets

Product Marketing

Products change significantly over the course of a market lifecycle. But how rapidly will products need to change? How significant will these changes need to be? Sometimes a company's product development roadmaps are ahead of market demand; sometimes they are behind. Either scenario can be used strategically. With the Market Maturity Matrix, the goal is to make this a conscious choice for you – so you know why your product timeline differs from the market, by how much, and how to anticipate transitions that may have dramatic impact on your product planning needs and efforts.

Product Planning	
Phase II • Typically 70-80% "standardized" product • Ability to customize to meet market-specific applications • Consistently meet target delivery dates (important for market perception, investors and sales) • Need interface or integration with complementary applications • Add formal training and education • Add formal documentation	**Phase I** • Typically 60-70% "standardized" technology or product • Ability to customize to meet unique customer requirements • Limited formal documentation
Phase III • Typically 80-90% "standardized" product or solution • Tight integration with complementary applications and legacy systems • Product or solution parity plus compelling differentiation • Consistently meet delivery dates • Supplemental consulting and aftermarket services	**Phase IV** • Typically 90+% "standardized" product or solution • Standards compliance • Feature/function "parity plus one" • Adopt or set industry standards

Pricing	
Phase II • Value-based • Pricing is established based on demonstrable value in creating competitive advantage • The more tailored the application to the specific need, the higher the value	**Phase I** • Technology-based • Experiment with pricing based on client needs, vision, and excitement level
Phase III • Competitive-based • Create competitive parity on price • Establish value based on unique capabilities and advantages • Can differentiate price parity based on speed and predictability of delivery and results	**Phase IV** • Cost plus • Create competitive parity on price • Establish value based on 'parity plus one' capabilities • Focus on operational efficiency to increase margin

Communications

Communications plays a critical role in internal and external messaging to all company stakeholders including employees, customers, prospects, press and media, analysts, investors, and shareholders. Communications is also central to establishing and maintaining company and service positioning as well as key brand elements. To do this, marketing communicators need to develop messaging frameworks that can be adapted across market phases. As a result, a company's value propositions, messaging and positioning statements, and brand attributes that encapsulate corporate vision may vary to reflect changing buyer characteristics.

Positioning	
Phase II • Domain expertise leadership • Focus on applications of technology • Focus on implementation proof, endorsement, and association (customers, technology and channel partners, investors)	**Phase I** • Technology leadership • Focus on building credibility • Focus on education • Highlight innovation, market lead, first mover, and "unfair advantage" • Emphasize key management and investors
Phase III • Customer service leadership • Complete solution leadership • Focus on competitive differentiation • Highlight service and support capabilities; customer satisfaction	**Phase IV** • Market leadership • Operational efficiency and cost leadership • Channel selection and availability leadership • Focus on brand values

Messaging (Why-to-buys)	
Phase II • Creates competitive advantage for customer • Key solution component • Vertical or segment specific expertise • Partnerships and alliances • Differentiation based on unique applications and use of product	**Phase I** • First-to-market • Technology advantage • Visionary appeal and opportunity • Discontinuous innovation
Phase III • Creates competitive parity for customer • Company reputation and leadership • Industry expertise • Customer success • Total solution • Support and service focus • Differentiation based on consistency, repetition, and measurement of the solution	**Phase IV** • Market parity • Brand value associations • Channel convenience and availability • Channel brand value • Service, selection, and price

Media, Analyst and Public Relations	
Phase II • PR focus on proof and implementation • Media focus on successful applications and customers • Analyst focus on market size and growth rate, competitive advantage, long-term strategy • Key tactics include success stories, speaking engagements, by-line articles, and announcing new customers and partners	**Phase I** • PR Focus on credibility and education • Media focus on visionary technology, leadership, large untapped market • Analyst focus on paradigm shift and inefficiencies of current solutions • Key tactics include speaking engagements and white papers
Phase III • PR focus on market expansion • Media focus on company growth, marquee customers, and market position • Analyst focus on market share, competitive advantage, barriers to entry for new competitors, and financial strength • Key tactics include company and product awards, customer testimonials, and announcing new innovations	**Phase IV** • PR focus on brand recognition • Media focus on market penetration into mainstream, channel penetration, and impact of technology on everyday life of users • Analyst focus on operational advantage, financial strength, and "what's next" • Key tactics include adherence to standards, parity plus one differentiation, announcing latest continuous innovation

Sales and Business Development

As markets mature, sales organizations, alliance partners, and channel partners need to evolve and adapt. This includes changes to the way sales organizations are structured and directed, the way sales and delivery channels are developed, and the way alliances and partners are pursued. In many organizations, account managers are responsible for customer needs and satisfaction. For this reason, customer relationship management is also included in this section.

Partnerships and Alliances	
Phase II • Alliance partners with complementary products that together create synergy for shared clients • Integration via Application Programming Interfaces (APIs)	**Phase I** • Strategic investment partners • Technology partners • Manufacturing partners • Supplier partners
Phase III • Strategic partners with complementary consulting services • Strategic partners who "complete" the solution	**Phase IV** • New technology partners (OEM suppliers) • Reseller partners

Sales Force Planning	
Phase II • Sales representatives with domain expertise in target market segments who understand business application and technical implications • Alliance and partner sales • Technology or product licensing and co-marketing	**Phase I** • Technical sales person with direct access to key engineers or developers • Sales engineers who intimately know the technology
Phase III • "Solution" sales representatives with executive focus • Emphasis on customer experience • Add incentives for account retention and satisfaction • Technical sales counter-parts • Channel sales or business development	**Phase IV** • Channel sales • Corporate account sales • Fewer technical sales counterparts

Selling Systems/Sales Tools	
Phase II • Sales demonstration and/or presentation • Application-specific customer case studies or success stories • Application white papers • Product ROI analysis • Customer profiles • Product brochures (application and/or segment focused)	**Phase I** • Technical sales demonstration • Technology white paper • Product data sheets • Product samples/trial versions
Phase III • "Solution" sales presentation • Generic customer case studies or success stories • Solution white papers • Topical white papers • Partner program kits • Corporate brochure	**Phase IV** • Point-Of-Purchase (POP) materials for retail • Co-marketing kits (partners)

Channel Development	
Phase II • Channel reseller partners with domain or application expertise in complementary market segments • Aftermarket development partners	**Phase I** • Original Equipment Manufacturer (OEM) and development partners – may require integration capabilities such as Software Development Kits (SDKs) • Develop Independent Software Vendors (ISVs)
Phase III • Channel reseller partners that can expand geographic or demographic reach • Systems Integrator partners with existing customer base and access to Fortune 500	**Phase IV** • Channel reseller partners at preferred customer points of purchase

Customer Relationship Management	
Phase II • Add user services such as training and education • Allow customers to interface with executives and product managers (customer advisory board) • Add user groups and programs • Involve customers in product definition, product test, and release process (Customer Involvement Workshops) • Personalized customer service by domain experts • Add communication programs (such as print and electronic newsletters, support updates, development updates, user conferences, etc.) • Customer co-marketing programs	**Phase I** • Share technology expertise • Demonstrate ability to customize to meet unique customer requirements • Allow customers to work directly with engineers • Demonstrate flexibility to meet individual customer needs • Involve customers in product design and definition phase (technology advisory board) • Introduce referral programs • Customer benchmarking
Phase III • Formalize support processes and training • Involve customers in product test and release process • Extend reach of solutions across other divisions or departments of client's business • Add programs to cross-sell and up-sell complementary products and services (increase share of customer wallet) • Add bi-directional communication vehicles (both inbound and outbound) via customer extranets, chat rooms, or special interest groups • Add channel partner programs	**Phase IV** • Least control of your customer • Registration programs • Extended warranties and guarantees • Automated service and support • Introduce programs to cross-sell complementary and non-complementary products • Extend extranet to have customer sections for sharing work products • Introduce loyalty programs

Operations

In marketing-driven organizations, the Market Maturity Matrix can be used to provide tactical considerations to other operations across the organization including support, development, finance, information technology, and human resources. For instance, it can provide insight into growth needs in human capital, financial capital needs for mergers and acquisitions, IT requirements to reflect changing market directions, and changes to improve how support services are delivered.

Operational Considerations	
Phase II • Ability to scale sales, development, production, delivery, and support • Ability to obtain resources knowledgeable in target market segments • Ability to add channels in specific market segments where domain expertise adds customer value • Ability to repeat existing successes	**Phase I** • Ability to obtain resources knowledgeable in technology • Ability to attract financial investment • Ability to sell unproven technology • Ability to produce, deliver, and support technology
Phase III • Ability to standardize product solution and implementation process • Add or expand channels in mainstream markets • Ability to create value for consulting services • Ability to streamline support services	**Phase IV** • Improved operational efficiency • Reduced development, production, packaging, and delivery costs • Improve channel services and efficiencies

MARKETING TACTICS BY MARKET PHASE: SERVICE LIFECYCLES

Market Development

As service lifecycles mature, each market phase offers unique opportunities to maximize market penetration and growth. This is significant for making key decisions related to market focus, selection, and expansion.

Market Selection and Expansion	
Phase III • Predominantly direct to targeted businesses within select market segments • Indirect through select resellers within complementary target market segments • Make approach participative to understand how service is being applied to business; turn into repeatable, packaged service • Use services to sell "products" (technology, content, etc.)	
Phase II • Integrated direct and indirect approach • Focus on leveraging methodology and service solutions to identify prospects who value your differentiation and approach • Prioritize outbound marketing by market segments or defined customer sets	**Phase I** • Approach is primarily indirect including viral marketing, networking, PR, etc. • Direct approach used to target marquee clients • Make approach interactive • Identify consistent customer needs to create repeatable delivery or management capabilities • Make approach educational (need to learn from every customer experience)

Market Focus	
Phase III • Targeted discrete market segments using service "packages" that leverage domain expertise or unique knowledge • Segment and focus on select target markets with custom packages • Selectively expand markets by adding domain expertise and replicating packages in new market segments • Form alliances and partnerships with resellers or technology companies with domain expertise or competence in target markets	
Phase II • Reinforce unique and defensible position with repeatable methodology or approach • Expand markets to serve customers of varied backgrounds and needs with consistent and predictable service solution • Co-marketing (alliances and partners) • Customer referral programs	**Phase I** • Target market segments based on identifiable, repeatable situations to early successes • Leverage channels for market entry in non-core segments • Identify and rank new, emerging, and unserved markets • Leverage technology source for lead referrals and co-marketing • Legacy product customers • Personal contacts • Purchased contracts or service agreements • Build unique and defensible position for leveraging service in unique way (industry expertise, pricing, consulting services, etc.)

Lead Development

Without priority and focus, lead generation programs can often look like a laundry list of activities. With the Market Maturity Matrix, marketers can prioritize and focus direct and indirect marketing activities such as trade shows, direct mail, telemarketing, on-line marketing, and advertising to align with the activities that typically yield the greatest return at that phase of the market lifecycle.

Trade Shows, Seminars and Conferences	
Phase III • Select shows and conferences targeted at core market segments • Application-driven educational seminars (either hosted alone or with partners)	
Phase II • Large industry shows and conferences targeted at buyers of alternative solutions • Technology/service source shows and events • Topical seminars (either hosted alone or with partners)	**Phase I** • Large industry shows and conferences • Educational seminars targeted at late adopters of technology or alternative solutions (buy versus build, dispelling technology myths, etc.)

Direct Marketing	
Phase III • Direct mail and telemarketing to prospects within target segments (either alone or with partners) • "Package" or application-focused messaging • Partner/channel programs to supplement target markets	
Phase II • Direct mail and telemarketing to prospects across broad range of markets • Solution-focused messaging (key on unique and defensible differentiation and position) • Participate in or sponsor technology/service source direct marketing programs • Partner programs to extend geographic reach and delivery options	**Phase I** • Participate in or sponsor technology/service source direct marketing programs • Leverage existing service providers with accessible customer base • Brand-focused messaging

Advertising	
Phase III • Select vertical or target market publications (alone or with partners) • Select technology industry trade publications • Selective targeted trade organizations (to reach target market segment) • Select publications targeted at channel partners • 'Package' or application-driven messaging	
Phase II • Technology industry trade publications • Select business/consumer publications • Target strategic solution buyers • Solution-focused messaging • Key on unique and defensible differentiation and position	**Phase I** • Heavily dependent on budget, revenue model and competition • Audit competitive advertising • Identify cost to create advertising parity with competitors • Select media channels likely to reach early service adopters

On-line Marketing	
Phase III • Targeted e-mail programs (ideally integrated with off-line programs) • Select banner ad placements and offers based on target market segment • Trade organization chat rooms and user group postings • Web seminars based on bundles or packages by target segment	
Phase II • Expanded search engine listings • E-mail based newsletter for customers • Cross-sell service solutions • On-line billing & account information • Web seminars based on industry solutions and customer experiences • On-line and self-service support • Automate methodology (via web, etc.)	**Phase I** • Parity with existing competitors and alternative solutions • Education-oriented (buy vs. build) • Search engine listings • Links to/from technology source • Links to/from partner sites

Service Creation and Development

Services change and expand significantly over the course of a market lifecycle. But how rapidly will new services need to be added? How significant will these changes need to be? Sometimes a company's services are ahead of market demand; sometimes they are behind. Either can be used strategically. With the Market Maturity Matrix, the goal is make this a conscious choice for you - so you know why your service delivery schedule differs from the market, by how much, and how to anticipate transitions that may have dramatic impact on your service planning needs and efforts.

Service Planning	
Phase III • Ability to customize service to meet market-specific applications • Ability to incorporate emerging technology into service delivery • Use services to sell "products"	
Phase II • Package or brand a unique methodology or approach (predictable and repeatable processes) for service delivery, implementation, and management • Expanded Service Level Agreements • Tight integration with complementary applications and existing systems • Integrate services to create service solutions (converge technologies or services from multiple partners) • Service solution "parity plus one" differentiation • Supplemental consulting and aftermarket services	**Phase I** • Access to technology or service source • Standards compliance • Ability to customize to meet unique customer requirements • Compatibility with existing client services and systems • Basic Service Level Agreements • Service economies of scale • Scaleable service infrastructure

<table>
<tr><th colspan="2">Pricing</th></tr>
<tr><td>Phase III
• Pricing is established based on demonstrable value in creating competitive advantage
• The more tailored the service to the solution, the higher the value</td><td></td></tr>
<tr><td>Phase II
• Create competitive parity on price through service value
• Establish value based on unique service solution capabilities and differentiation
• Create value based on predictability of service delivery and results</td><td>Phase I
• Create parity on price with technology purchase amortized over the life of product use
• Establish service value and differentiation based on brand
• Focus on operational efficiency to increase margin</td></tr>
</table>

Communications

Communications plays a critical role in internal and external messaging to all company stakeholders including employees, customers, prospects, press and media, analysts, investors, and shareholders. Communications is also central to establishing and maintaining company and service positioning as well as key brand elements such as voice. To do this, marketing communicators need to develop messaging frameworks that can be adapted across market phases. As a result, a company's value propositions, messaging and positioning statements, and brand attributes that encapsulate corporate vision may vary to reflect changing market needs.

Positioning	
Phase III • Domain expertise leadership • Focus on applications of service within target markets • Focus on implementation proof, endorsement, and association (customers, service partners, etc.)	
Phase II • Customer service leadership • Complete service solution or integration leadership • Focus on competitive differentiation • Highlight service consistency and predictability • Emphasize support capabilities and customer satisfaction	**Phase I** • Operational efficiency and cost leadership • Company, management and investor credibility • Brand values • Economies of scale and/or scope • Technology expertise

Messaging (Why-to-buys)	
Phase III • Creates competitive advantage for customer • Service solution component • Application or domain expertise • Partnerships and alliances • Differentiation based on unique applications and use of service	
Phase II • Creates competitive parity for customer • Company reputation and leadership • Customer success • Total service solution • Simplify and/or automate existing processes • Support and service focus • Differentiation based on consistency, repetition, and measurability of methodology or approach to service delivery or management	**Phase I** • Technology expertise • Economies of scale and/or scope • Financial advantages (cost savings, time reductions, resource reductions, etc.) • Convenience and availability

Media, Analyst and Public Relations	
Phase III • PR focus on proof and implementation success • Media focus on successful applications of service by customers • Analyst focus on market size and growth rate, competitive advantage, long-term strategy • Key tactics include success stories, speaking engagements, by-line articles, and announcing new customers and partners	
Phase II • PR focus on market expansion and competitive differentiation • Media focus on company growth, marquee customers, unique methodology, and market position • Analyst focus on market share, competitive advantage, methodology, barriers to entry for new competitors, financial strength • Key tactics include company and product awards, customer testimonials, and move towards service solutions	**Phase I** • PR focus on credibility, education and brand recognition • Media focus on buy versus build paradigm shift, service differentiation, and business viability • Analyst focus on operational advantage, financial strength, and 'what's next'

Sales and Business Development

The Market Maturity Matrix provides valuable insight for sales and business development planning. As markets mature, sales organizations, alliance partners, and channel partners need to evolve and adapt. This includes changes to the way sales organizations are structured and directed, the way sales and delivery channels are developed, as well as the way alliances and partners are pursued. In many organizations, account managers are responsible for customer needs and satisfaction. For this reason, customer relationship management is also included in this section.

Partnerships and Alliances	
Phase III • Partners with complementary services and products that together create custom applications for clients	
Phase II • Strategic partners with complementary services • Aftermarket development partners	**Phase I** • Strategic investment partners • Technology partners • Supplier partners • Sub-contract partners

Channel Development	
Phase III • Channel reseller partners with domain or application expertise	
Phase II • Reseller partners that can expand geographic or demographic reach • Systems Integrator partners with existing customer base and access to Fortune 500	**Phase I** • Reseller partners looking to re-sell into existing customer base • Service solution providers

Sales Force Planning	
Phase III • Sales representatives with domain expertise in target market segments who understand business application of services • Alliance and partner sales • Technology or service licensing and co-marketing	
Phase II • "Solution" sales representatives targeting executive audiences • Add incentives for account retention and satisfaction • Account managers for cross-sell and up-sell • Technical sales counterparts • Partner and channel acquisition	**Phase I** • Service solution sales reps targeting technical audiences • Corporate account sales • Technical sales counterparts • Executive networking referrals

Customer Relationship Management	
Phase III • Customize service to meet unique customer requirements • Personalized customer service by domain experts • Add communication programs (such as print and electronic newsletters, support updates, service updates, user conferences, etc.) • Customer co-marketing programs	
Phase II • Formalize support processes and training • Extend reach of service solutions across other divisions or departments of client's business • More aggressively cross-sell and up-sell complementary services (increase share of customer wallet) • Add bi-directional communication vehicles (both inbound and outbound) via customer extranets, chat rooms, or special interest groups • Add channel partner programs	**Phase I** • Grow service use from tactical value to strategic value • Service Level Agreements • Introduce programs to cross-sell complementary services • Introduce loyalty programs • Customer benchmarking • Introduce referral programs

Operations

In marketing-driven organizations, the Market Maturity Matrix can also be used to provide tactical considerations to other operations across the organization including support, development, finance, information technology, and human resources. For instance, it can provide insight into growth needs in human capital, financial capital needs for mergers and acquisitions, IT requirements to reflect changing market directions, and changes to support to improve how support services are delivered.

<table>
<tr><th colspan="2">Operational Considerations</th></tr>
<tr><td>Phase III<ul><li>Ability to obtain resources knowledgeable in applications of service in discrete market segments</li><li>Add channels in specific market segments where domain expertise adds customer value</li><li>Offer market-specific service packages or solutions that add strategic value to client</li></ul></td><td></td></tr>
<tr><td>Phase II<ul><li>Ability to standardize product solution and implementation process</li><li>Add or expand channels in mainstream markets</li><li>Improved customer service processes</li><li>Ability to scale sales, development, production, delivery, and support</li></ul></td><td>Phase I<ul><li>Ability to obtain resources knowledgeable in technology</li><li>Reduced development, production, packaging, and delivery costs</li><li>Improve services and efficiencies</li><li>Financing to create infrastructure necessary to result in economies of scale for customers</li></ul></td></tr>
</table>

Section III:
Self Assessment

12

You Are Here: Find Your Location On The Market Maturity Matrix

One of the most valuable uses of the Market Maturity Matrix is helping to identify where your market is right now. After all, it's hard to know how to get somewhere if you don't know where you're starting from.

Although determining market maturity using the Market Maturity Matrix is best done interactively using input from team members across an organization, the following self-assessments are the next best way. Included are separate Assessment Questionnaires for Product Lifecycles and Service Lifecycles

These questions are written with multiple choice answers in a variety of categories such as competitive landscape, existing marketing practices, and product delivery strategy. The answers are descriptive. Whenever possible, try and select the answer that best describes your current market, not your company.

PRODUCT LIFECYCLE
MARKET MATURITY ASSESSMENT

1. Which answer best describes the current level of industry awareness of your technology or product?

    1. Very few people have heard of our technology or our products.
    2. Our technology or product is known and recognized in small market segments or special interest groups.
    3. Our product and company are well recognized within our industry as well as broader markets.
    4. Our product and company have an established reputation both with consumers and/or our distribution channel.

2. Which answer best describes the applications associated with your technology or product?

    1. Our technology or product is used by a select group of visionaries to solve specific technical problems or used as a component in a larger technology solution.
    2. Our products are the core part of a solution that solves strategic business problems where its use resulted in a change in business process or efficiency.
    3. Our products are a strategic part of a company's operational efficiency or productivity and they play an important role in our customer's daily business.
    4. Our products are purchased more for financial gains or tactical improvements in business processes or efficiency.

3. Which answer best describes current demand for your technology or product?

    1. Very few people are yet aware of our technology or its benefits. We see this as latent demand where we need to convince people why they should try our product.
    2. Segments of the market have a real need for our product where it can result in new ways for them to compete or operate. However, only select individuals within these companies feel this need.

3. Multiple market segments have demand for our product or solution and it is packaged to satisfy the unique needs of each market segment.
4. Businesses everywhere have a need for our product or solution. It is standardized and packaged to appeal to mass markets.

4. Which answer best describes the price sensitivity of your current market?

   1. Because there are few choices available, people who know they need our technology or product are willing to pay a premium for it.
   2. Due to its unique features and capabilities, select individuals or markets will pay a premium for our technology or product, based on its ability to deliver them competitive advantage.
   3. Most buyers expect competitive pricing for our product versus similar offerings. Value for our product is often derived from our complementary services, experience, or brand recognition.
   4. Because our market is so competitive, it is increasingly difficult to create value through our product's capabilities; so we often use discounting strategies to close business.

5. Which answer best describes the preferred buying channels of your audience?

   1. People looking for our type of technology or product will buy direct from the manufacturer or developer.
   2. Most people looking for our type of technology or product will purchase either direct from the manufacturer or developer or from a preferred service or solution provider.
   3. Since most people buying our products are implementing a complete solution, they will either look to us or to full-service providers like system integrators to implement this solution.
   4. Most people purchase our products through our indirect channel partners, such as retailers, VARs, or system integrators.

6. Which answer best describes the motivation of your target audience or current customers?

    1. They are visionaries looking to be the first to have it.
    2. They are usually trying to gain a competitive advantage within their market space.
    3. They are usually trying to maintain parity with their competitors who have already adopted products similar to ours.
    4. They are usually trying to replace systems that are becoming obsolete or unserviceable.

7. Which answer best describes the primary business pain or business driver to purchase your technology or product?

    1. Buyers of our technology or product are very excited and emotional about it because it is new.
    2. Buyers of our products are usually trying to create newer, more efficient business processes or competitive advantage.
    3. Buyers of our products are usually trying to improve the efficiency of an existing process or operation, to automate it, or make it run smoother or faster.
    4. Buyers of our product or solution are usually driven by financial reasons. For instance, it is more cost effective to replace existing products than to maintain them.

8. Which answer best describes the risk tendencies of your primary audience today?

    1. Buyers of our technology or product are usually pretty risk prone. They are willing to accept risk to purchase, install, or use our product in order to be the first to have it.
    2. Buyers of our products will share the risk of implementing and using our product in order to achieve its benefits.
    3. Buyers of our product or solution are risk averse. They look to us (or one of our resellers) to provide them with a proven solution.
    4. Buyers of our product or solution are very conservative and typically avoid risk. Their decision to purchase usually stems from eliminating the risk of maintaining the solution they currently have.

9. Which answer best describes the level of involvement customers have in your product development process?

    1. Our customers are very involved in our ongoing development and specification process. They help us determine the structure, design, or content of our next version.
    2. Our customers help us validate our product development plans. We generally expose them to these plans during development or specification stages where we can refine our plans based on their input.
    3. We generally preview pre-production releases of our product to a select group of customers in order to gain their feedback and validation.
    4. We generally rely on customer feedback from technical support and surveys to influence features and functionality of future developments or versions.

10. Which answer best describes the trade shows and events attended by your target audience?

    1. Our target audience generally attends technology-oriented trade shows, looking for new technology innovations.
    2. Our target audience generally is involved in vertical or niche industry trade organizations and attends conferences, seminars and events sponsored by or related to their field of interest. Exhibits are usually a component of these events.
    3. Our target audience generally attends educationally focused executive or industry conferences and events, where their primary interest is education on industry trends and strategic business solutions. Exhibits are usually a component of these events.
    4. Our target audience usually attends large, technology industry trade shows, with the specific purpose of evaluating exhibits of companies with products and services to satisfy their needs.

11. Which answer best describes your current status with respect to partnerships and alliances?

    1. We have identified a few potential technology or business partners, but have not initiated any formal discussions.
    2. We are currently working with a select group of partners with complementary offerings. These partnerships were generally driven by specific customer needs involving sharing of technical information, collaborating on development, or creating links between our products.
    3. We are working with a number of partners in both development and distribution. Our development relationships are more established and our distribution/reseller partnerships are becoming increasingly strategic and are beginning to represent a significant part of our revenues.
    4. We have both technical and distribution partnerships that represent a significant revenue source for and impact on our business.

12. Which answer best describes your current handling of customer service requests?

    1. We have very informal customer support and service capability. The senior managers or the technicians and developers handle most of it.
    2. We have a person or a few people who have technical experience whose primary job is to support customer technical problems. Service is usually handled through the sales organization.
    3. We have people dedicated to servicing customer needs, whose job it is to be the customer's advocate within our organization.
    4. We have a formal support center and processes, where most people call for questions or technical assistance. Everyone in this center has access to the customer's account history and a technical support database for frequently asked questions.

13. What is the level of technical understanding of your products within your target audience?

    1. Our target audience has little technical understanding of our products and generally requires a high degree of education.
    2. Our target audience has a fair understanding of our technical capabilities and some understanding of the unique value of our product for solving specific technical problems in support of more strategic business needs.
    3. Our target audience understands our products and those of our competition well enough to understand the differentiation between these competitors' offerings and how these products or solutions can solve important business problems.
    4. Our audience does not want to know too much about the technology behind our products. Rather it wants to know the impact it will have on the bottom line.

14. Which answer best describes the technology used at the core of your product or solution?

    1. The technology we use for our products is very new or a paradigm shift from current products.
    2. Our product is a unique application for this technology, applied to a specific market niche or segment.
    3. Similar technology to ours is used in a variety of applications and solutions across a wide range of markets.
    4. The technology we use for our products is standardized. Everybody uses pretty much the same technology at the core of their product or solution.

15. Which answer best describes the level of competition in your marketplace?

    1. Because our technology or product is rather unique, we have little to no competition.
    2. There are many companies touting our type of technology. However, within our target market segment, we have only a few competitors, most of whom are small and niche oriented.
    3. Due to industry fall-out and consolidation, we compete with a few established competitors who have a presence across multiple markets and channels.

4. We are an established competitor in multiple markets and channels for our products. As well as our traditional competitors, we now face new competitors who are trying to commoditize the pricing and/or delivery to our target audience.

16. Which answer best describes the level of introduction of products in your market?

    1. No one in our industry has packaged the technology into a product yet for general consumption. Most are still flushing out final stages of development, manufacturing or assembly.
    2. Most competitors with similar technology or products to ours have released their initial versions and are touting their next versions.
    3. Most competitors with similar technology or products to ours have released multiple versions of their products, each representing incremental enhancements to the original product.
    4. All of our competitors have fully developed products that are packaged for general consumption and are sold through multiple channels.

17. Which answer best describes the level of standardization within your industry?

    1. Because our technology is still so new, there are no standards or guidelines in place within the industry for manufacturing, usability, quality or delivery.
    2. There are some standards emerging in our industry for base materials, components or application code, but each competitor's offering is still considerably different.
    3. Most of the major competitors use comparable technology, components, quality standards, or application code in their product or solution.
    4. Our industry has adopted standards for technology components, which are governed by a standards board, committee, consortium, or other entity.

18. Which answer best describes the merger and acquisition activity in your market?

    1. Because many of our competitors expect to ultimately dominate the market, we view our competition as the enemy, so there has been little merger or acquisition activity.
    2. Some of the companies in our market are beginning to merge or be acquired with competitors and complementary product providers. This consolidation is making the resulting companies technically stronger with larger customer bases.
    3. As a result of industry consolidations, mergers, and acquisitions, our remaining competitors are fairly well established. There are still some independents, but they focus on niche markets.
    4. Our market is very stable. The a few significant competitors remaining are all very big and unlikely to merge or be acquired by one another.

19. Which answer best describes the level to which business and technical analysts follow your industry?

    1. We work with a few technical analysts that understand our technology and plan to follow our market.
    2. We work closely with the technical analysts who follow our market and have received interest from the business market and financial analysts, who have taken an interest in our market.
    3. We work closely with both technical and business market analysts. Both are familiar with our company, our management, our industry and the market growth potential. The financial analysts rely primarily on the business and technical analysts for their information.
    4. We work more closely with the business and financial market analysts than we do the technical analysts. The technical analysts are asking about our next generation products; the business market and financial analysts what to know how long demand for our products will last.

20. Which answer best describes the development level of trade organizations or special interest groups in your industry?

    1. There are no formal organizations or special interest groups focused on our technology yet, although some existing groups have taken an interest.
    2. Many early users or adopters of our products have formed special interest groups, chat rooms, or associations to discuss the technology and its potential.
    3. The industry trade group for our technology and products has established an annual conference and/or trade show, prints a newsletter or magazine, and promotes industry gurus.
    4. Our industry trade association is involved in setting guidelines and standards for our technology and has established certification standards for experts in the field.

21. Which answer best describes the development process in your company?

    1. Our development process is very informal and centralized around our chief technologist. Our chief technologist often decides what goes in the products. Or we just sit in a room as a team and decide what goes into the products. We don't provide the engineers or developers with market specifications or detailed descriptions.
    2. Development is a little less centralized than it once was. We have begun to formalize our development process with regular meetings, a flow diagram of work to be done, and deadlines for each deliverable. Although most work is still done by a central group of people who simply divide up the work.
    3. Our development process is established and fairly predictable. We use specialists throughout the development process to streamline the workflow and perform specific functions.
    4. We have a documented development process that everyone understands. This process is constructed so that non-specialists can often perform the individual steps based on detailed specifications and instructions.

22. Which answer best describes the ownership of product development in your company?

    1. Our technology and products are developed and packaged by our engineering department.
    2. We have begun to incorporate more market feedback into our development process by converting this feedback into user specifications. Engineering then works to incorporate this into the development plan.
    3. Development relies on the direction of executive and marketing teams, based on customer and market demands for creating specifications, designing user interfaces and expanding platform or standards support. This direction is used to drive development timelines and the content of future releases.
    4. Marketing owns the development process. No development project is initiated without formalized market analysis and specifications.

23. Which answer best describes the competitiveness of sales situations in your industry?

    1. Competitors in our space rarely see one another because most customers want to move quickly and typically avoid competitive evaluations.
    2. Most of the competitors in our market have clearly identified market segments or positions that they specialize in. We often see the same competitors.
    3. Most of our major competitors market and sell in multiple segments and across multiple channels, so at times we are competing with new, unfamiliar competitors. We focus mainly on differentiating ourselves by our experience, our reputation and our customer satisfaction.
    4. Most of our major competitors are well established and recognized and market customers through a variety of channels (direct, VAR, retail, etc.), where they can capitalize on existing relationships.

24. Which answer best describes the condition of products in your market?

    1. Because products in our market are still very new, they are only about 50%-60% complete. Most are customized for each customer.
    2. Products in our market are probably about 70%-80% standardized. They are usually customized based on the unique customer needs and applications within certain markets.
    3. Products in our market are about 90% standardized. Although customers can often change configuration (feature/function selection, user interface choices, packaging choices), these are usually based on standards underneath. This makes these products good for VARs and system integrators who can customize solutions for their customers relatively easily.
    4. Products in our market are about 98% standardized. Our competitors have taken the most requested and popular user interface designs and features/functions and packaged this for the broadest audience. This also makes them more readily sold through mass channels.

Locating Product Lifecycle Market Maturity

Add up the individual point values of the answers to each question above. Then divide this total score by 24 = ________ (Product Market Maturity). This final number is a fairly accurate estimate of your Product Lifecycle Market Maturity.

To plot this on the Market Maturity Matrix, use Diagram R that indicates points along the product lifecycle curve. Draw a circle in the vicinity of your Product Lifecycle Market Maturity on the Product Lifecycle curve. For example, if your score is 2.6, you would draw a circle on the product lifecycle curve just below the horizontal line, in lower left Quadrant of the model.

Diagram R: Product Lifecycle Market Assessment

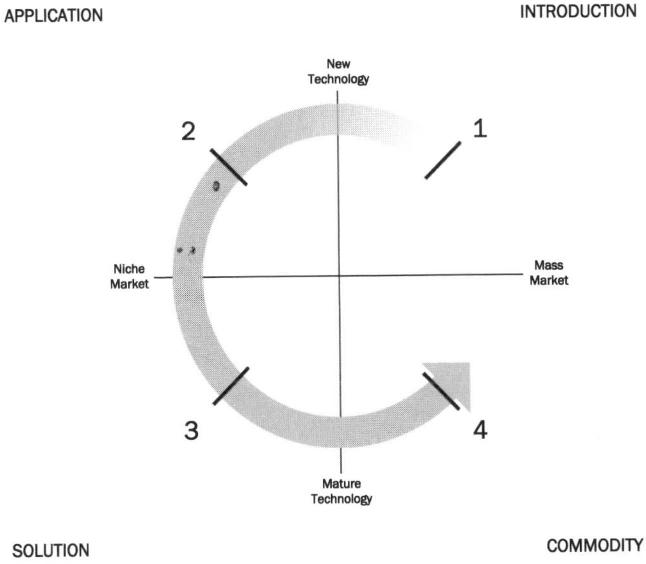

Notes:

____

____

____

SERVICE LIFECYCLE MARKET MATURITY ASSESSMENT

1. Which answer best describes the current level of industry awareness of your service?

    1. Our services are based on popular technology and applications, but our company does not yet have an established name and reputation within our industry.
    2. Our services and company name are recognized within our industry for our ability to add value to the underlying technology or applications we leverage.
    3. Our services are known and recognized in small market segments for leveraging the base technology or applications to create competitive advantage for our customers.

2. Which answer best describes the level of demand for your service?

    1. Many people use the underlying technology or applications our service is based on, but they have not yet accepted this technology as a service.
    2. Much of the market has a real need for our service, where it can result in new ways for them to operate.
    3. Select market segments have unique demands for our service that can result in competitive advantages for their business.

3. Which answer best describes the price sensitivity of your market?

    1. Many people buying our service are buying it for the short-term tactical gains and cost savings.
    2. Many people buying our service are buying it for strategic reasons, primarily based on improving business operations.
    3. Most people buying our service are buying it due to its unique application or advantage for their business and its ability to create competitive advantage for their business.

4. Which answer best describes the preferred buying channels of your audience?

    1. Buyers of our type of service are looking to purchase from a source that offers economy of scale and application expertise.
    2. Buyers of our type of service are looking for proven industry leaders or large system and service integrators with proven track records in this field.
    3. Buyers of our type of service are looking not only for someone with economy of scale and application expertise, but also someone who understands their business and industry well enough to see applications for the service beyond the basics.

5. Which answer best describes the motivation of your target audience?

    1. Financially motivated to reduce costs and or internal resources.
    2. Operationally motivated to improve business process efficiency, quality, or predictability.
    3. Motivated to create competitive advantage for their business within their industry or market.

6. Which answer best describes the risk tendencies of your primary audience today?

    1. Buyers of our service are very conservative and typically are motivated by risk avoidance. Their decision to purchase usually stems from more risk associated with maintaining the solution or service they already have.
    2. Buyers of our service are risk averse, particularly with implementation and ongoing management. Hence, they rely on us or one of our resellers to provide them with a proven, turnkey, or complete service solutions.
    3. Buyers of our service are willing to try something new or different with the service and share the risk of implementing and using our service in order to achieve greater value.

7. Which answer best describes your current level of customer interaction?

    1. We are very involved with our customer implementations to make sure that the service is delivered and used properly.
    2. We are active with each of our customers, particularly in finding ways for improving their operational efficiency and cross-selling and up-selling them on additional services.
    3. We are actively involved in the application of our service within our customers, particularly as it relates to continuing to finds ways to strategically leverage this service to give them competitive advantage.

8. Which answer best describes your current status with respect to partnerships and alliances?

    1. We have established relationships with key technology and product vendors that are the foundation of our service.
    2. We have expanded our partnerships beyond technology and product vendors to other complimentary service providers to create complete service solutions for our mutual customers.
    3. We have created partnerships with a few key industry or business specialists. They consult and aid us in implementing our services with customers where their area of expertise helps to extend the value of our service to help our customers create a competitive advantage for their business.

9. Which answer best describes the current level of competition in your marketplace?

    1. Although our service is new, it is not unique; and there are few barriers to entry. Hence our market is very competitive and already price sensitive.
    2. We have several direct and indirect competitors that offer comparable, but different services, or levels of service packaged such that they emphasize a unique approach or method for delivering or managing the service.
    3. Because we have focused on a couple of key industries or market segments, our services offer a unique value to our customers.

10. Which answer best describes the formalization and packaging of service delivery and management within your market?

    1. None of our competitors are currently touting a structured approach or methodology to delivering their services.
    2. Many of our competitors have introduced a methodology or solution approach to service delivery and management.
    3. Although many of our competitors have formalized methodologies and comprehensive service offerings, few have created service packages or bundles that create unique applications of the service for targeted audiences or market segments.

11. Which answer best describes the level to which business and technical analysts follow your industry?

    1. There are a number of technical and business analysts who follow the technology underlying services, but are not yet tracking or analyzing the service model market.
    2. We work closely with the technical and business analysts in our market, who have begun to create market sizing and forecast estimates for growth and key market vendors.
    3. We have branched out beyond general industry technical and business analysts to those industry analysts that cover the specific market segments we focus on.

12. Which answer best describes the service creation process in your company?

    1. Our service creation process is very informal. The founders or a small group of executives make decisions on service offerings based on perceptions of market and gut feelings.
    2. We have begun to formalize our service creation process, with regular meetings, a flow chart diagram of work to be done and specific goals for packaging and delivering these services. We base decisions on market trends and customer feedback.
    3. Our service creation process is established and fairly predictable. We use industry research and formalized ROI decision criteria to determine service viability.

13. Which answer best describes how you differentiate your services from your competitors?

    1. Our key differentiators are primarily determined by the capabilities of our underlying technology or product, the economy of scale of our service, and the expertise of our people in our field.
    2. Our key differentiators are based on our proven methodologies and processes for service delivery and management as well as our comprehensive service offering.
    3. Our differentiation is based on our expertise in certain industries, market segments, and applications of the service to help our customers create competitive advantage.

14. Which answer best describes the competitiveness of sales situations in your industry?

    1. Because many of our competitors use the same or similar underlying technology or products for their services, our competitors force us to be very price competitive.
    2. Because we have been able to differentiate our services based on our unique capabilities, delivery and management methodologies, or scope of services, we enjoy better price protection than we once did.
    3. Since we specialize in applying our services to specific market segments, we charge a premium for our service due to the high value our customer's place on our ability to leverage this service to create a competitive advantage for their business.

15. Which answer best describes the condition of services in your market?

    1. Services in our market are about based on standardized products and/or technologies. There are few barriers to entry.
    2. Although services in our market are based on standards, creating and delivering services at scale or using a formalized methodology are keys to differentiation and market expansion.
    3. Services in our market are standardized for the general market, but are fairly well customized for customers within our key target market segments. For these customers we often create unique implementations or service solutions to help them create competitive advantage.

Locating Your Service Lifecycle Market Maturity

Add up the individual point values of the answers to each question. Divide this total score by 15 = ________ (Service Market Maturity). This final number is a fairly accurate predictor of your Service Lifecycle Market Maturity. To plot this on the Market Maturity Matrix model, use Diagram S that indicates points along the Service Lifecycle curve. Draw a small circle in the vicinity of your Service Market Maturity. For example, if your score is 2.6, you could draw a small circle on the service lifecycle curve just above the horizontal line, in the upper left quadrant of the model.

Diagram S: Service Lifecycle Market Assessment

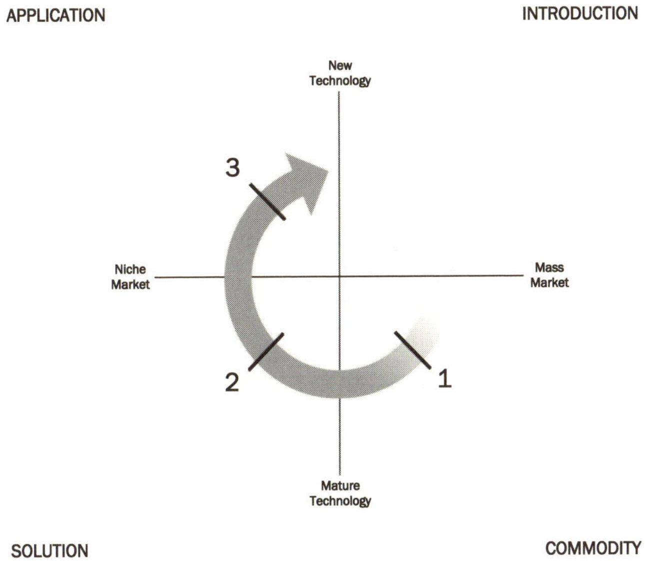

Notes:

___

___

___

13

The Next Move Is Yours

So you made it to the end of the book. We appreciate that. We also know you must be anxious to get out and apply what you have learned. So, we'll make this brief.

By now, we have hopefully convinced you that doing "the right marketing at the right time" is a little less daunting when you know what to look for… and when.

In fact, after using this approach, many people have come to us and said something to the effect of, "since we started using the Market Maturity Matrix, we can't remember how we approached market strategy before," or "this is the best approach to marketing I have ever seen." In essence, it is helping individuals and companies stimulate their thinking about market strategy and planning, while providing credible guidance for prioritizing resources and activities.

If you are beginning to feel the same way, we encourage you to take the next two important steps:

First, don't keep this book a secret. Recommend this book to colleagues and company executives, and require it for your service providers such as advertising agencies, public relations agencies, web design firms, media buyers, copywriters, etc. By sharing this with internal and external marketing teams, the lessons in this book can help you set a common frame of reference that can inspire everyone on your team think, act and make decisions faster and more strategically.

Second, use the Market Maturity Matrix as a basis for creating your own overlays and best practices for your company and your market. Challenge your marketing team to identify the specific characteristics and tactics unique to your market. In doing so, you can speed your planning and budgeting process, and improve resource prioritization and focus, even if your market undergoes dramatic changes. It could prove to be the best investment you make in your marketing this year.

With these two steps, we hope that "the right marketing, at the right time" becomes a mantra that offers a new secret weapon for creating competitive advantage in your market. We hope that by applying the Market Maturity Matrix to your business, you can succeed when others fail, standout in a crowded marketplace, and steer your business to market leadership.

Good luck and good marketing.

Index

H

I

J

K

L

M

N

O

P

R

S

T

U

V

W

About The Authors

Patrick Perugini has been in technology marketing for the last fourteen years. He has held senior marketing and product management positions with several leading international companies including Artemis International, Funk Software, Gores Technology Group, Wang Software, and Winchester Systems. Patrick is the creator of the Market Maturity Matrix™ and co-creator of the Marketing Scorecard™. Through his insight and perspective, he has earned the respect of industry executives and helped numerous prominent technology companies capitalize on emerging market opportunities. Patrick adds a new dimension to marketing's role in setting corporate strategy and helping companies focus and prioritize their resources and efforts. Patrick holds a Bachelor of Science degree in Marketing from Babson College.

Jeff Wiss is a 20-year veteran of technology marketing. He has successfully implemented marketing and sales programs for a wide variety of consumer and business-to-business products and services, as well as commercial World Wide Web sites. Jeff has been consulting to technology companies on market strategy, product and service introductions, and positioning and messaging for over 13 years, including work with such prominent companies as webMethods and Borland/Inprise. Jeff has also headed some of the most successful product launches in the technology industry, including the introduction of Dr. Solomon's Software, Rand McNally TripMaker, and The Far Side Computer Calendar. He brings a wealth of experience, insight, and focus to company and product positioning, messaging, market development, and channel planning.

Patrick Perugini and Jeff Wiss are co-founders of Maccel Group, a marketing acceleration company specializing in market planning and launch strategy for technology companies.

Using trademarked strategic marketing tools including the Market Maturity Matrix™ and the Marketing Scorecard™, the Maccel Group provides a disciplined approach to market strategy and tactical marketing planning. This innovative approach allows technology companies to better understand and predict changes in their market and devise business and marketing plans that are more credible, actionable, and measurable than ever before.